New Directions in Philosophy and Cognitive Science

Series Editors: **John Protevi**, Louisiana State University and **Michael Wheeler**, University of Stirling

This series brings together work that takes cognitive science in new directions. Hitherto, philosophical reflection on cognitive science – or perhaps better, philosophical contribution to the interdisciplinary field that is cognitive science – has for the most part come from philosophers with a commitment to a representationalist model of the mind.

However, as cognitive science continues to make advances, especially in its neuroscience and robotics aspects, there is growing discontent with the representationalism of traditional philosophical interpretations of cognition. Cognitive scientists and philosophers have turned to a variety of sources – phenomenology and dynamic systems theory foremost among them to date – to rethink cognition as the direction of the action of an embodied and affectively attuned organism embedded in its social world, a stance that sees representation as only one tool of cognition, and a derived one at that.

To foster this growing interest in rethinking traditional philosophical notions of cognition – using phenomenology, dynamic systems theory, and perhaps other approaches yet to be identified – we dedicate this series to "New Directions in Philosophy and Cognitive Science."

Titles include:

Robyn Bluhm, Anne Jaap Jacobson and Heidi Maibom (*editors*)
NEUROFEMINISM
Issues at the Intersection of Feminist Theory and Cognitive Science

Jesse Butler
RETHINKING INTROSPECTION
A Pluralist Approach to the First-Person Perspective

Massimiliano Cappuccio and Tom Froese (*editors*)
ENACTIVE COGNITION AT THE EDGE OF SENSE-MAKING
Making Sense of Non-sense

Matt Hayler
CHALLENGING THE PHENOMENA OF TECHNOLOGY
Embodiment, Expertise, and Evolved Knowledge

Anne Jaap Jacobson
KEEPING THE WORLD IN MIND
Mental Representations and the Sciences of the Mind

Julian Kiverstein and Michael Wheeler (*editors*)
HEIDEGGER AND COGNITIVE SCIENCE

Michelle Maiese
EMBODIMENT, EMOTION, AND COGNITION

Richard Menary
COGNITIVE INTEGRATION
Mind and Cognition Unbounded

Zdravko Radman (*editor*)
KNOWING WITHOUT THINKING
Mind, Action, Cognition and the Phenomenon of the Background

Matthew Ratcliffe
RETHINKING COMMONSENSE PSYCHOLOGY
A Critique of Folk Psychology, Theory of Mind and Simulation

Jay Schulkin (*editor*)
ACTION, PERCEPTION AND THE BRAIN
Adaptation and Cephalic Expression

Tibor Solymosi and John R. Shook (*editors*)
NEUROSCIENCE, NEUROPHILOSOPHY AND PRAGMATISM
Brains at Work with the World

Rex Welshon
NIETZSCHE'S DYNAMIC METAPSYCHOLOGY
This Uncanny Animal

Forthcoming titles:

Miranda Anderson
THE RENAISSANCE EXTENDED MIND

Maxime Doyon and Thiemo Breyer
NORMATIVITY IN PERCEPTION

New Directions in Philosophy and Cognitive Science
Series Standing Order ISBN 978–0–230–54935–7 Hardback
978–0–230–54936–4 Paperback
(*outside North America only*)

You can receive future titles in this series as they are published by placing a standing order. Please contact your bookseller or, in case of difficulty, write to us at the address below with your name and address, the title of the series and one of the ISBNs quoted above.

Customer Services Department, Macmillan Distribution Ltd, Houndmills, Basingstoke, Hampshire RG21 6XS, England.

Challenging the Phenomena of Technology

Embodiment, Expertise, and Evolved Knowledge

Matt Hayler
University of Birmingham, UK

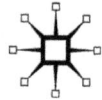

© Matt Hayler 2015

All rights reserved. No reproduction, copy or transmission of this publication may be made without written permission.

No portion of this publication may be reproduced, copied or transmitted save with written permission or in accordance with the provisions of the Copyright, Designs and Patents Act 1988, or under the terms of any licence permitting limited copying issued by the Copyright Licensing Agency, Saffron House, 6–10 Kirby Street, London EC1N 8TS.

Any person who does any unauthorized act in relation to this publication may be liable to criminal prosecution and civil claims for damages.

The author has asserted his right to be identified as the author of this work in accordance with the Copyright, Designs and Patents Act 1988.

First published 2015 by
PALGRAVE MACMILLAN

Palgrave Macmillan in the UK is an imprint of Macmillan Publishers Limited, registered in England, company number 785998, of Houndmills, Basingstoke, Hampshire RG21 6XS.

Palgrave Macmillan in the US is a division of St Martin's Press LLC,
175 Fifth Avenue, New York, NY 10010.

Palgrave Macmillan is the global academic imprint of the above companies and has companies and representatives throughout the world.

Palgrave® and Macmillan® are registered trademarks in the United States, the United Kingdom, Europe and other countries.

ISBN: 978–1–137–37785–2

This book is printed on paper suitable for recycling and made from fully managed and sustained forest sources. Logging, pulping and manufacturing processes are expected to conform to the environmental regulations of the country of origin.

A catalogue record for this book is available from the British Library.

A catalog record for this book is available from the Library of Congress.

For Regenia Gagnier, the most supportive mentor that I could have hoped for.

For Matthew Flanagan, for making me think better and at all.

For Richard Lee, my best friend and confidant, who always helps me tackle mountains.

For Briony Wickes, for taking a risk, taking my hand, and reminding me how much I like an adventure.

And for my parents, Brian and Jackie Hayler. There are no words, but all of these are for you.

Contents

Acknowledgements	viii
Introduction	1
1 *Fighting the Tools of Our Nature:* Technology in the Popular Imagination	7
2 *Beyond Common Sense:* Technology by Definition	60
3 *All Is One but Not for All:* Technology as an Object Encountered in the World	119
4 *Brushing Against Reality:* Technological Interactions Require Knowledge	164
5 *What Everything Knows:* Technologies as an Embodiment of Knowledge	207
Conclusion	230
Bibliography	235
Index	249

Acknowledgements

I started writing this book a long time ago at the University of Exeter. A huge thank you must first go to Flanny, Jos, and Bianca for making sure that the earliest drafts were written under the most ideal living conditions that I've ever had as an adult, full of great food, great company, and the best conversation; we need to hang out more. Carrie, thanks for all the Exeter dancing and for teaching me about poetry, see you in Brum soon. Ben and Lucy, you've always been the best hosts and walking companions, thank you.

Thanks also to the friends and colleagues there who spoke to me for eight years about various aspects of academic life and writing, took me out for dinner, drinks, and poker, and generally made my time there less difficult and a lot more fun: Andy Brown, John Dupre, Jana Funke, Regenia Gagnier, Felicity Gee, Gabriella Giannachi, Jason Hall, Siân Harris, Stephen Hinchliffe, Joe Kember, James Lyons, Robert Mack, Andrew McRae, Sinéad Moynihan, Alex Murray, Dan North, Sam North, Andrew Pickering, Vike Plock, Derek Ryan, Laura Salisbury, Philip Schwyzer, Florian Stadtler, Andrew Thorpe, Corinna Wagner, Paul Williams, Paul Young and Patricia Zakreski. Special mention must go to Sam Goodman, Victoria Bates, and Ryan Sweet who have become the greatest of friends as well as my academic heroes for their diligence, passion, and constant belief that the job, and thinking about things in general, can and should be fun – also V danced, Sam climbed, and Ryan trained me, all of which were always appreciated and anticipated. Thanks also to the Exeter Record Club, Nick, Jon, Gareth, and Ian, for taking me in, keeping me sane, and turning me on to all the best music.

My undergrads at Exeter made me better, challenging my thinking, making me laugh, and ensuring that I found out that teaching is the best job in the world. My PhD students, Richard Graham and Richard Carter, will go on to great things and are already writing important work. They've both talked with me about many of the ideas presented in this book – any mistakes here are errors that they would never have made. Zoe, thanks to you and Rich for all the calming drinks, burgers, chats, dinners, and support; your hard work is an inspiration.

The biggest change to my academic career has come from being part of the Cognitive Futures research network; I am forever indebted to Lisa Otty, Peter Garratt, and Mike Wheeler for their generosity and excellent company – this book and my job at Birmingham stem from my time working with them.

Thanks to Richard and Lorna Lee, Rob and Sarah Maythorne, and James Burnett, because you've all helped for ages, maybe without knowing just how much. James, you've been teaching me for forever, ta.

Jackie Hayler generously read through the chapters in their final order (thanks mum!), and Flanny, Jos, Ben, Sam, Rich, James, and Briony have been in endless discussions with me over the years about various sections – the mistakes are mine, any clarity is theirs.

A huge thank you to my anonymous reviewer and proofreader who made many helpful suggestions at the final stages. Thanks also to Esme and Brendan at Palgrave Macmillan and Vidhya at Newgen Knowledge Works for their support and expert marshalling of the final documents.

I wouldn't have thought about the issues discussed in this book if I hadn't read vital work by Davis Baird, Anthony Chemero, Andy Clark, Shaun Gallagher, James Gleick, Graham Harman, N. Katherine Hayles, Don Ihde, Matthew Kirschenbaum, Joseph C. Pitt and Henry Plotkin. Mark Z. Danielewski changed my life by changing the way I thought about reading over a decade ago with *House of Leaves*.

Introduction

I first began thinking about the ideas developed in this book around seven years ago. Amazon's Kindle e-reader had just been released and there were rumours that Apple was about to expand its revivified empire with the release of a new "slate" device, a tablet that would become the iPad. Working in an English department and writing about such things and talking about them at conferences, you regularly hear variations on the theme of the travesty of the end of the book. Responses range from denial, to the stoic, to the wailing and gnashing of teeth. People who've heard your papers have a tendency to send follow-up emails. There are often capital letters and multiple exclamation marks.

Over time I became more fascinated with the language that people were using to describe the new technologies of reading, the artefacts themselves, than the books that were meant to be read on them. I wanted to know if this language of resistance was new (it's not) and whether it could help us to understand the resistance to other kinds of technologies (and part of the argument of this work is that it can). So this book isn't really a book about reading on screen, though we'll come back to that move from the printed page's reflection of light to a digitally projected glow again and again. Rather, this is a book about the language that we use to talk about technology, and about what technology helps us all to do, how it can affect and facilitate our thought and action in the world, and how the bodies of our artefacts work in tandem with our own embodiment throughout the process of our lives. This will take us down some seemingly strange paths.

The first thing that is likely to come across here is the diversity of fields that I've been forced to draw upon. In trying to understand the nature of technology, and particularly of e-reading, one of our newest branches of artefact, I've had to find my evidence in Philosophy (phenomenology, postphenomenology, object-oriented ontology, epistemology), Evolutionary Biology, Cognitive Science, Literary and Bibliography Studies, the Digital Humanities, Archaeology, Anthropology, Science and Technology Studies, Sociology, and probably a few more depending on how you want to draw the disciplinary boundaries. I'm aware that this initially looks like I mess (or at least a set up for one) and I so I must ask, here and elsewhere, for the reader's patience and a little trust. I'm drawing on each of these disciplines not in order to pretend that I can contribute to them all, but instead to demonstrate that the limited claims of this work have both implications for and support from research far outside of the study of the equipment and practices of e-reading or of technology more broadly. And, despite this diversity, everything that I want to say, and near every writer that I cite, is concerned with embodiment and materiality, with the interactions between things in the world and their impact upon one another. For the vast majority of the discussion here, that concern will be turned to the kinds of activity that we call "use," the amateur or expert putting to work of our bodies in concert with some tool, some technology. I want to argue, however, that the use of technologies is at once complex, messy, and fundamental to what it means to be human and with this claim comes the realisation that technology cuts across and intrudes upon a huge swathe of academic and more quotidian concerns.[1] In order to understand technology we must understand its mess of relations, and by better understanding technology we also, in turn, open ourselves to better understanding how and what we know; how we act and how we *can* act; how we think and where we think it.

Phenomenological readings of e-reading devices are numerous, but some of the best aren't where we might expect them to be. The astute observations of the effects of these new devices on the body and haptically influenced psychology of their users are, as yet, under-represented in textual and digital studies research. Scholars such as Anne Mangen[2] are starting to explicitly address these issues from a Cognitive Science perspective, but these effects also need to be discussed by Humanities scholars who are invested in new forms of text and their position in media history. Instead,

---

[1] I don't mean to suggest that academic concerns are somehow loftier, rather that they don't always tend to be the day-to-day concerns of human lives. Something that I love about the philosophy of technology is its requirement that the academic cannot be separated from the everyday – the way we act each second becomes elevated, the most important and most rich object of concern, to which any pretence at intellectual rigour must visibly struggle to grasp in its intricacy.

[2] See, for example, *The Impact of Digital Reading on Immersive Fiction Reading*.

and particularly in the absence of longitudinal empirical work, the great abundance of interesting commentary on this subject is to be found in popular reports, blog posts, reviews, comments, editorials, and the like, and it's this living archive that often acts as a provocation to the chapters here – *these* are the voices demanding that we pay attention to what's going on and in its full richness, we just need to recognise the value in listening.

In the same way that folk psychology has become recognised as a source of evidence,[3] so we can see the various reports of e-reading's unnaturalness and ergonomic inadequacies as a form of "folk phenomenology," a description of experience stemming from first person analysis which the reporter often feels can be exported, with limited modification, to other experiencers of the same or similar phenomena. Thomas Metzinger, in one of the few available classifications of the term, describes folk phenomenology as "a naïve, prescientific way of speaking about the contents of our own minds – folk-phenomenology is a way of referring specifically to the contents of conscious experience, as experienced from the first-person perspective...and is characterized by an almost all-pervading naïve realism." As folk psychology can often demonstrate useful examples, methods, and states to its academic counterpart, I would like to argue that folk phenomenological intuitive report, despite its significant issues as a source of insight, also has a lot to offer in terms of prompting us towards the concerns that are central to negotiating what is qualitatively different about reading on a portable screen. Because beyond the interest, beyond the enthusiastic sales figures and adoption rates, digital reading has prompted resistance from bibliophiles, academics, high-school teachers, bloggers, parents, journalists, book-shop owners, librarians, and myriad other commentators from all areas of public discourse.[4] Their voices represent an often

---

[3] For an overview of the promise and problems with folk-psychology, see Andy Clark "Folk Psychology, Thought, and Context."

[4] For a selection across this diversity, see, for example, Sven Birkerts *The Gutenberg Elegies: The Fate of Reading in an Electronic Age*; Robert Coover "The End of Books"; Caleb Crain "Twilight of the Books"; Alan Kaufman "The Electronic Book Burning"; Mark Ruxin "The Death of Touch and the Lost Joy of the Unexpected"; Christine Shaw Roome "I've Got the Screen Eyes to Prove It: How Do Ebooks Really Compare to Traditional Books"; Sarah Schofield "Ten Things I Hate About eReaders"; Ben Ehrenreich "The Death of the Book"; Shane Richmond "The Printed Book Is Doomed: Here's Why"; Sam Leith "Is This the End For Books?"; David Dobbs "Is Page Reading Different From Screen Reading?"; Susan Greenfield "We Are at Risk of Losing Our Imagination"; Gary Frost "Reading by Hand"; and Max Bruinsma "Watching, Formerly Reading."

surprisingly coherent position, one unmatched in voracity or constancy by e-reading's supporters. Such coherence emerges from somewhere deeper than the affordances of voguish new equipment, tapping into some of our most fundamental concerns with being in the world in tandem with the tools that we devise – though they must often be tied to the particular artefacts of the day, I want to position these concerns as something enduring in themselves and to try and do justice to them; they are the reasons I was sent to the body, brain, and to daily practice. The academic literature in the field must continue to fail us while it plays catch up – it's not that we don't have the tools, but we are still working out exactly where to place them, and my suggestion here is that folk phenomenological report gives us a much needed approximate map.

The discussion here is divided into five chapters that each approach the issue of technology from a different angle while trying to build up a consistent terminology born out of the issues raised. In Chapter 1, I consider the discourse of resistance surrounding e-reading and how it reflects a rhetorical attitude that is repeatedly played out in the history of the introduction of new technologies. As we will see, new artefacts, particularly those designed for mass adoption or in specialist areas of the arts, are often described as "unnatural," with such critique tending to be arrayed around two distinct threads: technology separates us from the world and technology corrupts our minds. I first outline how these threads play out in a wide variety of discussions about e-reading and then begin to challenge the assumptions, demonstrating that a printed book is no more "natural" than an e-reader, and that using an e-reader might be seen as a perfectly "natural" thing for human beings. This forms the basis for the claim that central to what it means to be human is our adaptation to and coevolution with tools, what Bernard Steigler describes as "technogenesis."

Chapter 2 considers how we define technology, first looking at the range of existing definitions and what we might understand as a "common-sense" definition of how the term is deployed in daily use, before going on to look at some of the significant failures of the existing approaches (not least that they allow for us to conceive of technology as something somehow autonomous or inhuman). The majority of the chapter is then taken up with a new definition of technology that sees it not as a class of objects, but as a class of phenomenological experience with several consistent features. I will argue that in order for an interaction with any equipment to be considered as technological it must Extend the user's abilities; exist in a Community of users; be Incorporated into the user; and Domesticate the user over time. The chapter concludes with a discussion of the implications of this definition including that only experienced users, experts, can achieve technological interactions; amateurs experience mere "devices," things for accomplishing a task, but not the rich encounters that accompany true technologies.

Chapter 3 considers how artefacts appear in our perception and how expert and amateur encounters with the same artefact reveal how our perceptions in general are shaped by experience. The chapter begins by looking at how multicomponent objects tend to be considered as singular, coherent things, even as more and more of their components are revealed to whoever encounters them. The construction of these "gestalt" objects is then discussed, initially with regard to the effects of metaphor, language, and community storytelling on how they appear, but then in terms of the embodiment of both user and artefact. The role of the affordances of the object are considered, followed by the role of the user's own embodiment in their cognition as revealed by contemporary Cognitive Science. These concerns can be allied with the field of postphenomenology, which explores the role that technologies play in mediating human experience of the world, and this gives us a circle where our embodied cognition and storytelling shape the way that we perceive our technologies whilst those same technologies impact upon our cognition and the stories that we are able to tell.

Chapter 4 aims to establish artefacts as embodiments of knowledge by developing the idea that experts and amateurs encounter different phenomenal objects when using the same real artefact. I outline some of the relevant features of Graham Harman's object-oriented ontology (OOO), uniting his concerns with the appearance and withdrawal of objects in perception with the Cognitive Science and postphenomenology of Chapter 3. Though I work extensively with Harman's ideas, this chapter isn't about advocating for the OOO project, but instead demonstrates how its insights into the tendency for objects to perpetually escape from our perception allows for artefacts to consistently surprise even the most expert of users. That experts are surprised less often than amateurs, however, suggests that there must be some increased correlation between the expert and the expertly deployed technology, and the second half of this chapter considers such a correlation and its implications for what it means to possess "knowledge." By moving the site of knowledge (in line with the perspectives of postphenomenology and embodied cognition outlined in Chapter 3) from the human brain to a cognising system that includes the body and artefacts in the world, we can start to understand human knowing as a subset of a more fundamental, posthuman mechanism for acquiring and deploying data. I describe knowledge at the end of this chapter as a growing coherence between any two (or more) systems, as the possibility of repeatable successful action, and as the slowing of the flow of information from the world. None of these claims rely on any human cognising agent – this is a view of knowledge and information that allows for the artefacts themselves to possess a knowledge of their users, a knowledge that facilitates increasingly expert use.

Chapter 5 considers how artefacts come to embody this knowledge and information. I first outline how we might consider the development of technologies as an evolutionary process and then explore the field of evolutionary epistemology and a stance that sees biological adaptations as instantiations of knowledge about the world. If technologies can be seen as having evolved, and if such an evolutionary epistemology stands, then there is further support for a posthuman conception of knowledge that can include nonhuman knowers, and our artefacts can therefore be seen as occupying a flattened position in technological interactions, contributing, if not equally, then comparably, to any human expert.

This is a book about the objects that are closest to us and their continual ability to surprise. It's a book about skill and knowledge, about how the strange relationships between humans and artefacts enable incredible and incredibly subtle actions, and about how those relationships are less asymmetrical than we might suppose. To state it clearly: if nothing else, by the end of this work I hope to have persuaded the reader that it may be productive to consider a technology as being a particular, expert phenomenological encounter with an artefact that also tends to possess its own knowledge of its environment of use. But this is a way away. We must start with the resistance – why can the use of equipment feel so wrong, and wouldn't we be better off getting our hands dirty?

1
Fighting the Tools of Our Nature: Technology in the Popular Imagination

In this chapter we will look at some of the distinctive attitudes people take towards new technologies and how such feelings often cut sharply against the grain of how technology is actually deployed in human life. One of the significant questions of this book is "how do we define 'technology'?", and in the second chapter I'll consider the key scholarly attempts to establish the term, but I can't begin to wrangle with those ideas before acknowledging the more emotive side of the argument. In our received notions of what technology is, there is a strong discourse of resisting new devices and defining technology by what it is not, principally by saying that it is not "natural." As we'll see, such ideas can manifest in strange ways – technology is a human thing, that's what makes it unnatural, and yet it is rarely seen as a part of our nature, at least until a new technology comes along to replace something successful in an existing field, in which case the older (and preferred) technology often tends to be positioned as now being the more natural, or more attuned to our being. Discussions of the new devices for e-reading will provide us with useful demonstrations of just these kinds of conflict, pointing us to the most vital issues for users involved in the early moments of accommodating new daily practices. But the same argumentative shape also repeats in a variety of cultural arenas, a sample of which will be explored below. I offer this diversity of descriptions of resistance not on the assumption that the plural of "anecdote" is, eventually, "evidence," but rather to jog the reader's memory: I assume that these kinds of argument are familiar, from the media, from family and friends, from books read and films watched, conversations overheard and participated in – I simply want to demonstrate their pervasiveness and reiterate some of their more distinctive manifestations: technology in general isn't natural; new technologies aren't natural; technology makes us stupid; technology keeps us apart from the world. There are also two loose but important shapes to be found in these negative definitions: technology is unnatural, it separates us from the world, and technology corrupts, it does something to our minds.

I'll argue later that these discourses are, at least partially, the result of an insufficient definition of "technology." But, as is easily seen in the arguments around e-reading, there are always legitimate concerns with the introduction of new, often more complicated devices – we always lose something. It is part of the task of this book to work through how people adapt to the new despite its challenges, but the counter-argument to the blunter resistant discourses that we'll encounter in this first chapter is simple: technology is natural in as much as it is a part of human nature. Much of the rest of the argument presented here rests on this fundamental fact, and as such it needs to be demonstrated ahead of time. But by presenting it in the face of what I expect to be very familiar arguments to the contrary, I hope to show that we can nuance the debate around technology whilst retaining the important concerns of those who might justifiably resist new equipment in any area.

By using that word, "resistance," I fully intend to invoke a political, moral, or ethical claim to avoiding or repudiating the move toward new technologies or new norms of use and, in the particular case of e-reading, to allowing a generation to grow up reading from screens rather than paper pages. Such stances range throughout the popular discourses surrounding the resistance to digital reading devices, from the essayist Sven Birkerts' assertion in *The Gutenberg Elegies* that "language and not technology is the true evolutionary miracle" (6), to David Gelernter's solution for digital technology being useful, but codex reading being somehow "right": "I assume that technology will soon start moving in the *natural* direction: integrating chips into books, not vice versa" ("The Book Made Better," my emphasis).[1] Ideas of rightness, essence, of what is correct are at the heart of much contemporary discussion of reading, and Birkerts is amongst the most eloquent detractors, picking up on this language of the unnatural throughout his work. His playing up of a dichotomy between the reader's "natural" interaction with a bound book and the "unnatural" processes of reading on a multimedia screen recurs frequently throughout the *Elegies*: "Running the eyes down column-inches of print is part of the former way of processing the world, but it is no longer the natural mode for many.

---

[1] Gelernter, a professor of Computer Science, would like to see codices augmented with certain digital elements; his examples include making it "beep" if you've misplaced it, or being able to search its text online, whilst maintaining the codex's functionality if the electronics fail.

Not when bits of information stream in from every source, there to be isolated and studied as needed" (238). It is clear from Birkerts' phrasing here that this new mode may have become the default, but to him it seems far from natural. This way of thinking reaches its apotheosis in his claim that: "[w]hat [codex] reading does, ultimately, is keep alive the dangerous and exhilarating idea that a life is not a sequence of lived moments, but a destiny. That God or no God, life has a unitary pattern inscribed within it" (85).[2] This final quotation gets us to the crux of the resistance to "unnatural" new reading practices: for those set most firmly against digital reading technology, codex reading, and all of its related practices, has become an almost religious or spiritual experience, tapping into something at our core. Fully bearing this out, the novelist Alan Kaufman, in a ferocious article entitled "The Electronic Book Burning," says of his hatred of digitisation:

> My books have been hard-won. What made it all seem worthwhile was the book, the physical item, a kind of sacred and appropriate temple for the text contained within. Had I been told from youth that my literary destination would be some 7 inch plastic gizmo containing my texts shuffling alongside thousands of other "texts" I would have spit in the face of such a profession and become instead a hit man or a rabbi... To me, the book is one of life's most sacred objects, a torah, a testament, something not only worth living for but as shown in Ray Bradbury's *Fahrenheit 451*, something that is even worth dying for... I have given the days of my life, my years, my youth and adulthood to the book, as both sacred object and text.

Not all commentators would go this far, of course, and I suspect that many of those who imply or deploy that word "unnatural" would deride the full extent of Birkerts and Kaufman's polemic. But it does feel as if each instance stems from a common pool, and part of the function of this chapter will be to investigate why the rhetoric of the unnatural is so pervasive in expressions of the mundane experience of new technologies as sources of frustration and concern.

---

[2] Italo Calvino describes a similar fear of the screen's usurping this fought-for order when he wrote in the early days of digital text: "you are gripped by the fear of having... passed over to 'the other side' and of having lost that privileged relationship with books which is peculiar to the reader: the ability to consider what is written as something finished and definitive, to which there is nothing to be added, from which there is nothing to be removed" (*If On a Winter's Night A Traveller* 112).

1.1 Technology is unnatural

The history of resistance to what we might instinctively call "technology" is most often associated with the start of the industrial revolution and the Luddite movement that saw outbreaks of violent dissent targeting the new machines of agriculture and production, and occasionally their owners, in the wake of widespread unemployment[3]. The historian Richard Bulliet, however, traces an even earlier history to the medieval Middle East in "Determinism and Pre-Industrial Technology." Bulliet questions the notion of inevitable and linear progression in technological development, his research demonstrating that communities resisted adopting certain efficient new technologies such as alternative yoking mechanisms for draft animals, the use of wheeled transport and wheelbarrows, and even early forms of print, all of which were deployed by other geographically proximal cultures. The reasons he cites for such refusal include class-based resistance, but also ethnic and lifestyle factors, attitudes of "us and them" with relation to surrounding cultures which prevented the adoption of outwardly superior new tools. It is easy to see how such moralism might manifest as resistance: "*they* use x, *we're* superior to them and we use y, therefore y is correct/right/natural" (a more persuasive discourse, perhaps, than "I'd rather suffer than be like them," and one with a ring of Nietzschean truth with regards to the origin of certain kinds of morality).

The reasons to resist technology, then, are numerous and have become deeply rooted around the world, significantly predating the proliferation of complex artefacts which seem to mark out various contemporary cultures as being somehow more technologically minded (more on this later). From the Enlightenment onwards, however, romanticised (and Romantic) resistances appealing to a return to natural living become increasingly widespread in direct proportion to the perceived impact of technology on the average citizen's daily life. The Romanticism scholar Steven E. Jones identifies a movement, "Neo-Luddism" (*Against Technology* 20), which draws upon the historical Luddite cause, if in bastardised fashion, as a source of utopic reasoning with regard to technology. Neo-Luddism can be taken to encompass viewpoints as diverse as a continuation of the original Luddite fear of displacing human labour

---

[3] For further discussion of resistance to technology and Luddism see Steven E. Jones *Against Technology: From the Luddites to Neo-Luddism* and Nichols Fox *Against the Machine: The Hidden Luddite Tradition in Literature, Art, and Individual Lives*.

with mechanical apparatus, to a full blown "technophobia," a more general fear of the negative potential inherent in increasingly complex, intrusive, or embedded technologies. But Jones notes a significant tension between modern Neo-Luddite resistance to (primarily digital) technologies and the original Luddite cause:

> Today "Luddite" often means "deluded technophobe"...[D]etermined weavers and cloth finishers, skilled artisans demanding fair wages and control over their own trade, [are] often wrongly interpreted as champions of the simple life and of nature, as voluntary primitives and Romantics...[But it was the Luddites'] right to *their* technology [that] they fought to protect, not some Romantic idyll in an imagined pretechnological nature. (3 & 9)

Rarely is the popular invocation of Luddism about a fight for workers' rights, instead appealing to an idyllic pretechnological state from which modern society seems to exponentially propel us. I am certainly not suggesting that contemporary fears are unfounded, indeed the crises of anthropogenic climate change, oil spills, extreme water shortages, grain speculations, crop monocultures, and antibiotic resistant superbugs are continual reminders that questioning the unrelenting pursuit of a nebulous "progress" is essential to any hope of a responsible consumer society. Jones, however, also describes a problem with uniform resistance: "There are undoubtedly real technology-based conspiracies or patterns of connectedness at work behind the scenes and a general suspicion in this regard is not paranoid in the clinical sense: It is prudent citizenship... [T]he nature of [a] paranoid response is a tendency to universalize its fears" (176). The homogenisation of resistant discourses around a natural vs. unnatural binary exemplifies such universalised fears and demonstrates that the questioning of technological change may also prove detrimental if left unquestioned itself. Screen reading, for instance, might mark a potential ecological improvement over the current printed book industry;[4] may allow for greater and more democratic access to cultural works from all literate cultures;[5] might encourage a generation

---

[4] A round up of blog posts and newspaper articles on this discussion can be found at the *eco libris* site ("ebooks vs paper books"). The consensus seems to be cautiously in favour with one sizeable caveat: don't upgrade your e-reader too regularly.

[5] See for instance the Worldreader and One Laptop Per Child projects that provide libraries of digital texts to developing countries including work by indigenous authors that had previously received extremely limited distribution.

used to television and computers not to abandon reading;[6] and may even simply be a convenient way to access written material. If the resistance to it is just a slavish adherence to an older trend of cynicism in the face of the new, then, no matter how useful or practical or vital that trend, it deserves to be interrogated, not least because it has much to tell us about both our attitudes towards new artefacts and our misunderstanding of how we actually deploy material culture.

1.1.1 Satirising scares

The resistance to modern technology, particularly scare stories surrounding the effects of digitisation, have become so widespread that they have begun to be analysed and lampooned in light of their historical counterparts.[7] Kathrin Passig, for instance, identifies some "Commonplaces of Technological Critique," arguments which have been put forth recurrently over the last century as generic criticisms of new equipment including:

- "What the hell is it good for?" – A phrase IBM engineer Robert Lloyd asked of the microprocessor in 1968.
- "Who wants it anyway?" – Harry M. Warner, one quarter of the Warner Brothers studio's founding family team, is said to have asked in 1927 "Who the hell wants to hear actors talk?"
- It's destroying our minds: – "People read...what is true and what is false mingled together, without examination, and they do this purely out of curiosity, with no real thirst for knowledge...Idleness becomes a habit and creates, as does all idleness, a relaxation of the soul's energy," this a warning from the *Universal Lexicon of Upbringing and Teaching* in 1844.
- If the new technology has to do with thinking, writing, or reading, then it will most certainly change our techniques of thinking, writing, or reading for the worse – "For critics around 1870, the postcard sounded the death knell for the culture of letter writing, while in February 1897 the American Newspaper Publishers Association discussed whether 'typewriters lower the literary grade of work done by reporters.'"

---

[6] See Lauren Barack "The Kindles Are Coming" report for the School Library Journal which discusses children's increased enthusiasm for reading on electronic devices for reasons of portability, secrecy (keeping how much reading they're doing from friends who might judge the activity), and novelty.

[7] Vaughn Bell provides a short history of scares surrounding new media including writing, radio, and mandatory education in his article "Don't Touch That Dial! A History of Media Technology Scares, From the Printing Press to Facebook." See also, Adrienne LaFrance, "In 1858, People Said the Telegraph Was 'Too Fast for the Truth'."

And Randall Munroe, writer of the popular webcomic *XKCD*, compiled a similarly revealing list of common queries about new devices:

THE SIMPLE ANSWERS
TO THE QUESTIONS THAT GET ASKED ABOUT EVERY NEW TECHNOLOGY:

WILL [___] MAKE US ALL GENIUSES?	NO
WILL [___] MAKE US ALL MORONS?	NO
WILL [___] DESTROY WHOLE INDUSTRIES?	YES
WILL [___] MAKE US MORE EMPATHETIC?	NO
WILL [___] MAKE US LESS CARING?	NO
WILL TEENS USE [___] FOR SEX?	YES
WERE THEY GOING TO HAVE SEX ANYWAY?	YES
WILL [___] DESTROY MUSIC?	NO
WILL [___] DESTROY ART?	NO
BUT CAN'T WE GO BACK TO A TIME WHEN—	NO
WILL [___] BRING ABOUT WORLD PEACE?	NO
WILL [___] CAUSE WIDESPREAD ALIENATION BY CREATING A WORLD OF EMPTY EXPERIENCES?	WE WERE ALREADY ALIENATED

"Simple Answers"

What's significant about these instances, what makes them funny, is their painful familiarity. Such ideas recur again and again in discussions of new technology, implying both a general fear of change, but also a common source or set of common sources. A history of railing against new technologies, and often in predictable fashion, contextualises current debates about e-reading, exemplifies Jones' universalisation of fears, and raises two important points: (i) that people frequently consider a move away from the equipment that they are used to using as being at the very least problematic, and at worst an outright threat, and (ii) when technologies have been around for a while they can start to seem to be a part of the natural order – "it's not these older technologies we have to worry about, it's those new ones that pose a threat, that separate us from the world."

1.1.2 Separation from nature

Nature is often seen as the privileged term in a dyad with culture, artifice, or the human, but this certainly hasn't always been the case, particularly in the West: "[m]uch of the extravagant hope generated by the Enlightenment project derived from a trust in the virtually limitless expansion of new knowledge of – and thus enhanced power over – nature" (Marx, "Postmodern Pessimism" 239). It is a resistance to this very project of dominion which shapes much Neo-Luddite critique, seen at an extreme in the extension of Rousseau's "noble savage" (the most typically Romantic of resistances) to the anti-technology, anti-civilisation "anarcho-primitivism" touted by the philosopher and self-professed Neo-Luddite John Zerzan[8] and his sympathies with the manifesto of Theodore Kaczynski, the Unabomber[9]. More moderate resistant voices do, of course, exist; the philosopher John Gray's masterful *Straw Dogs*, for example, questions the modern faith in the progress of technology as a replacement for religious belief, and earlier, in the American Romanticism of Henry David Thoreau, Ralph Waldo Emerson, and William Carlos Williams, we can sense a very similar thread:

> Men have become tools of their tools. (Thoreau 61)

> Here are great arts and little men. Here is greatness begotten of paltriness... Every victory over matter ought to recommend to man the worth of his nature. But now one wonders who did all this good... 'Tis too plain that with the material power the moral progress has not kept pace. It appears that we have not made a judicious investment. Works and days were offered us, and we took works. (Emerson "Works and Days")

> Machines were not so much to save time as to save dignity that fears the animate touch. It is miraculous the energy that goes into inventions here. Do you know that it now takes just ten minutes to put a

---

[8] "It seems to me we're in a barren, impoverished, technicized place and that these characteristics are interrelated" (Zerzan, "Against Technology" 1).

[9] See Zerzan's *Elements of Refusal*, *Future Primitive*, and *Running on Emptiness*, and Kaczynski's "Unabomber Manifesto" (originally titled *Industrial Society and Its Future*).

bushel of wheat on the market from planting to selling, whereas it took three hours in our colonial days? That's striking. It must have been a tremendous force that would do that. That force is fear that robs the emotions: a mechanism to increase the gap between touch and thing, not to have contact (Williams 177).

In the work of each of these writers there is a sense of the diminishment of human dignity, significance, and independence to artificial attachments. Again we see a moral concern and again we see the two distinct and Cartesian threads that I'm trying to establish here: technology separates the body from the world and it corrupts the mind.

In their work on the history of music, Trevor Pinch and Karin Bijsterveld note the way that such concerns around new technology getting "in the way" appeared at the introduction of new musical instruments – the pianoforte, for instance, was "for some an unwarranted intrusion by a mechanical device" into a "musical culture that revered the harpsichord" (537). The introduction of "linked-key mechanisms and valves" onto woodwind instruments also "met with opposition, because they ruled out the possibility of making a 'vibrato by simply moving the fingers over the sound holes' and diminished the player's ability to 'correct out of tune sounds' by means of slightly altering finger positions. One commentator...declared that improving tone quality by the use of keys was 'neither complex nor art'" (539)[10,11]. New technologies, threatening old ways, are again seen as blunting human interaction – ease is suspect.

The philosopher of technology Andrew Feenberg goes as far as to define technology in light of its capacity for mediation:

> God creates the world without suffering any recoil, side effects, or blowback. This is the ultimate practical hierarchy establishing a one-to-one relation between actor and object. But we are not gods...Technical action represents a partial escape from the human condition. We call an action "technical" when the actor's impact on the object is out of all proportion to the return feedback affecting the actor. We hurtle two tons of metal down the freeway while sitting in comfort listening to Mozart or the Beatles. (48)

---

[10] Citing Christian Ahrens "Technological Innovations in Nineteenth-Century Instrument Making and Their Consequences" and Hubert Henkel, "Die Technik der Musikinstrumentenherstellung am Beispiel des klassischen Instrumentariums."

[11] See also Karin Bijsterveld "'A Servile Imitation': Disputes about Machines in Music, 1910–1930."

Technology allows us to temporarily exert a dominion usually beyond us, diminishing our perceived effort whilst maximising our output and drawing us closer to the work of a deity immune to Newton's third law, hampered only by our persisting embodiment. But the price, it seems, is its acting as a barrier, as an escape from human nature rather than an accordance with it, barricading us from and inoculating us to the realities of the world.

And this isn't necessarily an aspect of a peculiarly modern technology; it's an aspect of tool use that goes back thousands of years. What *is* different may simply be the extent to which such "visceral insulation" is deployed in near every aspect of modern life. This is a theme of the work of the archaeologist Timothy Taylor (whose thesis of unavoidable human entanglement with technology in *The Artificial Ape* will become significant very shortly). Taylor fully realises the extent of mediation inherent in tool use from the outset, describing it as a "reverb of increasing technology": "the technology that insulates us from cold and exhaustion also insulates us from the psychic rawness of nature. Visceral insulation is the trend toward disengagement from the actuality of hunting, killing, and gutting, or, in the case of domesticated animals, of rearing for the table and then physically dispatching" (*Artificial* 98). Undoubtedly, modern technology is connected to a distancing of the harsher realities of keeping a large population fed, clothed, and sheltered, but it's not new. Taylor positions visceral insulation, as I would like to, as part of the fundamental nature of technology; we can (and will) nuance this, but for now it's enough to note that technologies do get between us and our world in various ways (all of which I'll refer to as visceral insulation, not just our being inured from the harm that our style of living produces) and that such mediation will always have a psychological effect.[12]

The phenomenological philosopher Martin Heidegger is often identified as being anti-technology in his stance against modern forms, particularly in "The Question Concerning Technology" where he conceives of the essence of contemporary technology as comprising a restricted "unveiling" of the world. We'll return to "The Question..." in the next chapter, but perhaps the most vivid example of comparing the old and the new in that essay is his juxtaposition of the windmill and the hydroelectric dam: the windmill works within and alongside nature, whereas

---

[12] To pre-empt the argument to come, which will manifest in various ways in each chapter, there is no neat, unmediated access to the world to be interrupted, and no reason to think that technological mediations should be deprivileged because of the specific mediations that they facilitate, particularly in light of their fundamental relationship to what it means to be human.

the dam tames it, reducing the river to the single purpose of producing electricity. Heidegger is most often read as seeing our technological encounters as an impoverished view of the world, one where nature is revealed simply in its having a use value, as being a "standing reserve" of materials (trees and coal to burn, seas of fish for food, rivers for power, etc.), a means to an end and nothing in itself.[13] This too can be seen as a form of visceral insulation, a keeping of us from the world "as it is," ensuring that any faith in our domination is misguided as we only tame a certain aspect of the environment in which we act.[14]

Kevin Kelly, in his quasi-spiritual, quasi-biological description of technology in *What Technology Wants*, describes his feeling of escaping the insulating effects of modern artefacts during a trip around Asia after dropping out of college: "My personal possessions totalled a sleeping bag, a change of clothes, a penknife, and some cameras. Living close to the land, I experienced the immediacy that opens up when the buffer of technology is removed" (1). "Immediacy" and "buffer" are the key terms here, a clear expression of the "purer" experience that can occur in the (relative) absence of technological intervention.

Again, I suspect that this is all at least a broadly familiar discourse: "in the old days people knew where their food came from, they were connected to nature, whereas now we have industrial farming and plastic packaging which distance us from the lively stuff of growing, rearing, reaping, and slaughtering." All true, of course, but as Taylor points out, this industrial technological intervention is far from modern. Finds at archaeological sites investigating late-fourth-millennium Mesopotamia often include thousands of rudimentary artefacts called BRBs, bevel-rimmed bowls, which Taylor compares to Styrofoam containers. Typically used by workers and slaves engaged in corvée labour producing large public structures, the BRB was probably filled only once with a ration of barley porridge before being discarded once its purpose was served.

What the BRBs mean is that we know that there were by this time people in the world who had lost the direct chain of contact with

---

[13] For a good overview of Heidegger's attitudes toward technology in this vein see Ronald Godzinski Jr. "(En)Framing Heidegger's Philosophy of Technology."

[14] Heidegger, however, knew the importance of technology in day-to-day life and, though it certainly concerned him, his approach is more nuanced than a simple "anti-technology" screed: "For all of us, the arrangements, devices, and machinery of technology are to a greater or lesser extent indispensable. It would be foolish to attack technology blindly. It would be shortsighted to condemn it as the work of the devil" (*Discourse on Thinking* 53).

the source of food that they ate...The psychological effect of BRBs should not be underplayed...The appearance of this type of mass-produced fast-food vessel was part of an intensifying retreat from the wild, and an ever-greater control over the terms of death. (*Artificial* 98–100)

Technology, then, has long been associated with distancing its users in various ways from their environment and experience, and this might well contribute to certain Neo-Luddite positions. It is unclear, however, what pristine past is being harkened to, what time where contact with the earth went unmediated (it certainly wasn't recent). This is an idea we must continue to pick up on.

So far, we've seen a little of the historical arguments for the old over the new, the natural over the unnatural, how technology is responsible for the distancing of human experience from the world and the corruption of the mind. Let's focus in now on how these repeating concerns can be found playing out in the specific example of the debates surrounding the more recent technological mediation of e-reading.

1.2   Resisting e-reading

In the chapters to come I want to talk about technology in a way that *does* refer to our natures, and *does* refer to the natural world, while also avoiding a Romantic appeal to a sense of the natural as simple, morally right, or anthropologically normal.[15] I've so far argued here, however, that much of the discourse around resisting technologies, and particularly new technologies in existing fields, frequently takes on a language appealing to just such a view of naturalness. We've seen variations on the theme, technology as corrupting influence, technology as mediating layer, technology as allied with the wrong society, etc., but I think that we can make the strong claim that these stances can be reduced to a binary of natural vs. unnatural, where "natural" is always a linguistic construct, a nebulous appeal to a superior default. The ways in which these arguments are formulated largely depends on the particularities of the technology of course, the specific materiality and effects, but there

---

[15] Perhaps the most influential critique of the "natural" in the late twentieth century is Derrida's discussion of Rousseau in *Of Grammatology*. In the early twenty-first century, a contender has to be Timothy Morton's *Ecology Without Nature* which explicitly attacks the romanticisation of a knowable nature in favour of an unknowable and posthuman "dark ecology."

seems to be a clear common thread. To reinforce this further, and to fully establish the discourse of the natural as a dominant, if not *the* discourse of the rejection of new technologies, I would like to spend some time looking at specific examples of popular, scholarly, and personal rejections of e-reading technology, to establish these tropes in a case study that we can return to, and to set up the importance of embodiment in understanding the nature of technology and its effects (an embodiment not just of the user, but also of the artefact deployed). The change in embodiment of the book, from a bound sheaf of paper pages to an electronic screen which, almost universally, is capable of projecting more than written script, raises its own particular concerns beyond the rhetoric of "codex reading is natural, e-reading is not." These issues, born out of the real experience of use, flavour the discussion, but the broad appeals identified above also remain. The first, to be discussed in detail below, is often positioned as the more romantic: digital technology for reading provides a layer of visceral insulation between us and the sensual world of which printed books are a very pleasant part; the codex is better suited to our natures as tactile beings. The second appeal, discussed in 1.2.2, is predominantly coded as a scientific concern: e-reading negatively impacts upon our cognition and impoverishes the experience of reading in a quantifiable way, this in comparison to a codex reading which, again, demonstrably better fits our natures. Again, e-reading establishes the discussion to come; the debates about its nature and effects model concerns about the nature and effects of technologies as a broad class of artefact.

1.2.1 E-reading and visceral insulation

In the quotation from William Carlos Williams above, the new agricultural technologies emerging at the turn of the twentieth century had separated workers and consumers from the world in order to save "dignity"; they represented a move away from intimate interaction for the sake of an unnatural propriety. There's a similar sense of this kind of mediation in Birkerts and Kaufman's work also cited above, that the new reading technologies separate us from a rich physical world of which the bound book is a unique part. Elements of this position are also widespread online (our largest data-set of folk phenomenological reports) with innumerable blog posts on the pitfalls of screen reading appealing to a perceived separation from material existence. The following is typical: "eReaders give nothing away about the journey you've taken with a book. No dog-ears, smells or smudges. They don't express a life shared with a story. In the future, people won't lovingly pass on the battered books they read repeatedly as kids" (Schofield, "Ten Things I Hate About eReaders").[16] Printed books, in their particular form, are seen as able to record in their materiality a history of use which ties them to the physical world in a way that the clinical asceticism of plastic and glass wholly fails to. I'm not concerned with refuting this point, at least not as yet, but what's important is that many users of e-readers see their devices as occupying a materiality that is distinct from their own relationship with the physical world. We live in a world of history, and wear and tear, and scents, and messiness, and memory, and therefore an object that clearly aligns with that world also clearly aligns with us. It's easy to see why we might, therefore, associate codices with an anthropocentric and evocative naturalness; e-readers are different (this is what makes them feel like technology).

N. Katherine Hayles, an influential critic and theorist of digital texts, argues against the abandonment of printed materials, suggesting an example of such embodied superiority: "the [printed] book [possesses] robustness and reliability beyond the wildest dreams of a software designer. Whereas computers struggle to remain viable for a decade, books maintain backward compatibility for hundreds of years" ("Print Is Flat" 84). Physical books, in this way, conform to our bodies and minds as other objects in our extended history have: they don't require power; they're always "on"; they're relatively robust; and their workings are intuitive with respect to our daily embodied experience. The same cannot be said of e-readers. As we'll see later, a heightened complexity of basic use doesn't rule out adoption and familiarisation, but we can intuitively understand initiates to e-reading baulking at the switch.

This is a phenomenological issue, a matter of the perception of an experience and its nature, which is to say that these kinds of questions about reading emerge from our existence as embodied beings who feel a familiarity or affinity in grappling with similarly physical artefacts. The philosophers Shaun Gallagher and Dan Zahavi, in their discussion of Cognitive Science and embodiment, highlight the importance of experience as being "bodily meaningful":

> To be situated in the world means not simply to be located someplace in a physical environment, but to be in rapport with circumstances that are

---

[16] Scofield's post is a good example of commonly identified problems with the new reading equipment emphasised in popular resistance, particularly perceptions of fragility ("an e-reader will break if my cat sits on it"), expense (Scofield invokes an imaginary phone call to a bus depot to negotiate an e-reader's safe return versus the sad, but relatively unproblematic loss of a codex), and battery life ("[t]hey rely on a battery which is bad for the environment, and impractical for camping").

bodily meaningful...Those possibilities that my body enables..., just as much as those activities that my body prevents or limits, and that define what is possible or impossible – these are aspects of embodiment that I live with, and through, and that define the environment as situations of meaning and circumstances for action. (*Phenomenological Mind* 137–138)

We define and appreciate our environment, and our equipment, in relation to our bodily positions and possibilities; as the phenomenologist Maurice Merleau-Ponty suggested in *Phenomenology of Perception*, being embodied is the condition for our having a world to experience (169).[17] We are also primed to encounter similarly tangible things in our environment: our radical dependence on our own bodies leaves us unhappy, particularly without training, in negotiating abstractions (e.g., probabilities, mathematics divorced from practice, correlations that aren't causes, etc.). This goes some way to explaining the response to the felt dimension of the e-reader experience: the union of book and object (i.e., that the book read and the codex held appear phenomenologically identical) may be a more effective, and seemingly natural, textual instantiation than the changeable e-reader and e-book theatre. In a codex, the book, the work, and the medium are as physical as one another, they are the same thing, bound in an unchanging dialogue within that one item, and we've grown to understand that this is the reading (and writing) experience: essentially, to acquire a specific and unique (though broadly replicable) object and to try and work out what it means to whatever standard we deem appropriate. The e-reader/e-book relationship is entirely different: "An electronic text literally does not exist if it is not generated by the appropriate hardware running the appropriate

---

[17] In this regard, and to which Gallagher and Zahavi are alluding in the above quotation, the originator of modern phenomenology, Edmund Husserl, also characterised the body
  as being present in any perceptual experience as the zero point, as the indexical 'here' in relation to which the object is oriented. It is the center around which and in relation to which (egocentric) space unfolds itself (Hua 11/298, 4/159, 9/392). Husserl consequently argues that the body is a condition of the possibility for the perception of and interaction with spatial objects (Hua 14/540), and that every worldly experience is mediated by and made possible by our embodiment (Hua 6/220, 4/56, 5/124). (Zahavi, *Husserl* 98–99)
The abbreviation "Hua" refers to the relevant volume/page number of the 34 volume collected writings of Husserl, the "Husserliana" editions. Zahavi's familiarity with this immense body of work in the original German ensures his usefulness to anyone interested in Husserl's thought, and his work in *Husserl's Phenomenology* has been essential to supporting my own understanding of Husserl necessarily drawn from English translations.

software. Rigorously speaking, an electronic text is a process rather than an object, although objects (like hardware and software) are required to produce it" (Hayles, "Print Is Flat" 79). The e-reader is as physical as any codex, although it establishes different gestures and actions during use, but the e-books that can be read from it are ephemeral, ghostly, temporarily brought to the surface of the tangible object before returning to somewhere else, leaving the physical form of the equipment to mean by itself and in other contexts, other combinations.[18] We have other corollaries for this experience of course, in television, computing, theatre stages, cinema screens, and varieties of music players: each of these are stable physical objects which can call up diverse and transient content. But perhaps it is therefore to be expected that when we make the obvious comparisons with these media, rather than with the printed book, a reader might ask "is this even reading anymore?" As Anne Mangen, a specialist in digital reading research, describes the experience,

> digital texts are ontologically intangible and detached from the physical and mechanical dimension of their material support, namely, the computer or e-[reader]...When reading digital texts, our haptic interaction with the text is experienced as taking place at an indeterminate distance from the actual text, whereas when reading print text we are physically and phenomenologically (and literally) in touch with the material substrate of the text itself. ("Hypertext..." 405)

This perceived silencing of the hands' unique chatter with the brain, preceded by the text's chatter with the hands, may be part of what motivates resistance to e-reading built around a lack of feel and the apparent promotion of intellectual impediment. As we appear to take our hands out of reading, when we remove our tactile observation, it seems to induce a particular kind of blindness. Representations and fears of blindness,[19] for all their ability to shock us with our own fragility, hold none of the horror of a true loss of touch, not just a numbness of the

---

[18] Again, the claim here is phenomenological rather than ontological. As Matthew Kirschenbaum ably demonstrates in *Mechanisms*, forensic techniques such as electron microscopy reveal that data doesn't disappear, and each bit (the smallest unit of computable information) possesses a unique physicality on the surface of a hard drive. But, perceptually, the text appears from and returns to nowhere even as the object persists in our hands.

[19] See, for instance, Derrida's study on representations of blindness in visual art, *Memoirs of the Blind*, or Freud's famous connection of the fear of blindness with the fear of castration described in "The Uncanny."

hands, but a removal of the skin from our sensation. To touch is never in our control – we touch against our will, – always maintaining at least a point in pressure with something, hence the fascination with acrobatics, zero-gravity, or the weightlessness of floating in a heavily-salted sea (though none of these represent a true loss of touch or else they would become grotesque). Never in our control, but for the most part controlled (pain can be excessive touching, or the echo of a misplaced touch). No wonder that so many avid readers, so many holders of printed books, feel that they must speak out: in the seeming intangibility of the text might they subconsciously fear that the visceral insulation of the new technology will make us, if not paralysed, then haptically blind? The period of adjustment to what is a very new tactile experience for an otherwise familiar activity has to bridge the drama of such profound worries.

Most significant for us here, however, is that to suppose intangibility at any level of the digital text is simply a misreading: the software has a physical forensic materiality at the level of the hard disk image and in the materiality that it entails from production to distribution to use, and it is instantiated on a device which is equally physical and entailed. That the phenomenological experience is of a discomforting immateriality is born of a misrecognition rather than a valid ontological claim. Practice and education, the notion of expertise that will become central to the rest of this book, will surely demonstrate that though the hands' chatter has changed, the chatter still occurs. What can seem alienating at first, however, is that it is not only use that has changed, but what and how that use can mean.

The historian Lucien X. Polastron, in *The Great Digitization and the Quest to Know Everything*, suggests that "the sole difference a paper book carries – in addition to the clearly superior epidermal pleasure it provides over that produced by touching plastic...is that the total weight of the text is constantly felt by the reader. This sensation perhaps gives the reader an impression...of possessing the whole of its meaning, an illusion whose loss could panic fragile souls" (35). This idea starts to close in on the motivation for some of the folk phenomenological reports of resistance to reading on-screen rooted in the tactile experience of the artefacts of reading: a paper book has come to *represent* knowledge, rather than just contain it, and its fixed and physical coherence, completed and separated from the world by its covers, projects the illusion of a definitive truth. Reading had always, until the advent of the moving image, meant interacting with an object which *is* the codex (or the scroll, or the tablet). Cinema, television, and computing changed that arrangement, and the establishment of e-reading threatens, if nothing else, to make that shift irrevocable. There is something to be lost here, though it might well be trivial, at least in terms of its importance to future generations of readers: with screen reading the book and the object are taken apart; any assemblage of the work and the substrate will be temporary.

On this point I'm grateful to Tim Carmody[20] for his comments on an early blog-post of elements of this chapter. There, Carmody invoked Gerard Genette's *The Work of Art: Immanence and Transcendence* as a way of negotiating this seeming separation of form and content:

> I'm all about this materialist-phenomenological approach to reading. But I think you may slight the way in which reading a book has always been a complicated interplay of immanence & transcendence... For instance, the codex book has never been the material FACT of the work of art the way that a sculpture or painting is. Likewise, the physicality of reading the paper codex is harder to ignore now that we have a very different (and on its face, less robust) physicality for reading all kinds of documents, including books... [W]hen you're working through the genuinely phenomenological (as opposed to the narrowly empirical) account of reading a book, its transcendence, the fact that it does not appear to be merely confined to that physical codex, is a genuine part of that experience. ("Reports...")

This is an important point. The *immanent* instantiation of the printed text (to use Genette's term) is essential to understanding the artwork, but, however much it conditions it, it is not the sole site of the work. The *transcendent* text extends away over every edition, and every edition's history, and historical conditions, and means of production, etc., in short into the realms of Book History, Bibliography, and Textual Studies. I'm certainly not trying to refute this idea, and Genette's distinction is elegant, but despite a reinvigorated interest in materiality after digitisation, and despite the transcendent spread of the artwork always already being present in printed works, it remains important to note that the folk phenomenological discourse demonstrates that a significant change has occurred with the *seeming* breakup of the text down the lines that I have described, even as the reader retains an awareness of a broader

---

[20] Carmody currently writes for a variety of publications on and offline, but previously worked as an academic specialising in reading and phenomenology. For more on Carmody's work on these issues see his "Immanence and Transcendence in New Media."

transcendent text which exists beyond the unique immanent object. I am not doing the codex a disservice in saying that it *often* seems to be the whole phenomenological fact of the text in use by a typical reader; this is one of its wonders, its seeming to capture an extended phenomenon between two pieces of cardboard, and a wonder that doesn't translate in satisfying fashion to the e-book + e-reader experience. It is a wonder that is lost in the transition, one that we should rightly consider the implications of, and it marks a subtle variant of e-reading's seeming to insulate readers from meaningful physical experience.

Lynne Truss, in her punctuation pedant's handbook *Eats Shoots and Leaves*, offers an illustrative folk phenomenological experience which supports this attitude: "Scrolling documents is the opposite of reading: your eyes remain static, while the material flows past" (181). Now, I don't agree with Truss' claim here, that the eyes don't move during reading when the material is scrolled rather than paginated; indeed, all physiological data about eye movement during any form of reading would run counter to it.[21] But this doesn't negate value in the report – infinite scrolling (such as can be found in some iPad reading apps (Wattpad for instance), on websites, or in desktop documents), as opposed to pagination, to Truss doesn't feel like reading at all, and in fact acts somehow as its opposite, the eyes seeming to do no work and the experience feeling passive in comparison to the overpowering trajectory of the machinery.

Christine Shaw Roome, a professional fundraiser for an academic library in Canada, reports a similar experience to Truss, writing at the blog *Life as a Human* about her first experience of reading from an iPad. She wonders if she's now even reading a book at all, so drastically has the feel of the activity altered:

> [Her husband interrupts her] "I'm reading a book!" But, was I? I was missing the tactile features of the book, which often comfort me. The smell and feel of the book and the way you can see how far you've read by measuring the thickness of the pages. When I buy a book, I always take time to look at its design – the typeface, the page weight and colour, the way the ends appear to be torn or are cut precisely. The texture of the cover and the photography or illustration that accompanies the title all draw me in and are part of the experience of enjoying a book. Sometimes, I buy a book just because I like how it feels in my hands.

---

[21] See, for example, Stanislas Dehaene's *Reading in the Brain* (13–18).

Roome offers us a good survey here of the most familiar elements of the folk phenomenological debate surrounding reading on screen: it no longer seeming to be a book, it not *feeling* like a book, it not smelling like a book,[22] the wedge of remaining printed pages acting as a consistent indicator of progress,[23] and the object as aesthetic artefact.

Anna Dorfman, while writing on Jonathan Safran Foer's experimental novel *Tree of Codes*, also argues for what is important to her interactions with print:

> I don't see the act of reading as a purely word-based experience. Reading is also tactile. Reading should involve interaction between you and the text in your hands. The speed at which you turn to the next page (or flip back to the one before) matters. That accidental glimpse you got of page 273 (while still only on page 32) while fishing around for your bookmark matters. The weight of the book in your bag – that subtle reminder that it's waiting for you – matters. The paper stock matters! The font, the letter-spacing, the margin width! It *all* matters! ... And don't even get me started on the smell of old paper and fresh ink!

What really comes through in these detailed experiences, besides the recurrence of olfactory satisfaction, is the importance of the haptic interaction; the feel of the book in the hands is an important part of grounding the experience as what it is. When this goes missing the effect

---

[22] The scent of physical books, old and new, is a frequently recurring issue in online and offline debates and one which has become such a shorthand for the deprivations of reading on screen that a spoof range of aerosols (SmellofBooks.com) did the rounds in various discussions of the subject. See for example Charlie Sorrell's "New Book Smell" at *Wired*'s Gadget Lab blog, Alison Flood's post at the *Guardian* book blog, "Making Scents Out of Novels," or the following: "Me, I've parted with most of my print library. For good. Ninety percent of my reading now takes place on-screen, although I'm uneasy about digital books living inside those intangible walled gardens. Can I pass them on to my kids, like my mother did with Camus to me? Will they keep my side notes? Will they smell?" (Ferro-Thomsen "Reading Beyond Words"). The appeal to smell might seem an odd reason to cling to a medium, but, if nothing else, it shows how deeply passions run in this regard, or just how far appeals will go to demonstrate the sanctity of the old form – everything about it is comforting.

[23] A fact of the reading experience that Jane Austen refers to towards the end of the playful *Northanger Abbey*: "The anxiety...can hardly extend, I fear, to the bosom of my readers, who will see in the tell-tale compression of the pages before them that we are all hastening together to perfect felicity" (185).

is so profound, the cognitive dissonance so great, that seemingly unintuitive questions, such as those seen in Roome's account, arise: "Is this even a book?", "Is this reading?" A last folk report further exemplifies this – the designer and critic Max Bruinsma describes a friend's worries that her new interactions with writing somehow don't "count":

> "I don't read," someone I know well told me. She meant that she doesn't read the way "readers" read. People who can spend hours on end with a book in a chair or on the sofa, occasionally turning over a paper page and appearing to have completely forgotten that there exists a world outside the sentences they are reading. No, she's not one of those readers. But, I say, you actually read the whole day through! You scan articles and books, browse through websites and online fora, open and answer emails, gloss over newspaper headlines. Yes, but that's not reading, she says.

These four examples give us a sense of the broad consensus around the disruptive potential of changes in technological form. The medium changes, so the activity changes, so the gestures change, and the sensual experience becomes interrupted; the effects are both nebulous and disconcerting.

A related and similarly uncanny reason for the sense of loss produced by the seeming division of work and matter is that the physical book, in its relatively uncomplicated materiality, resonates with our own bodies in a way that the e-reader as magic box cannot. Karen Littau, in *Theories of Reading: Books, Bodies, and Bibliomania*, argues that "[t]he relation a reader has to a book is also a relation between two bodies: one made of paper and ink, the other flesh and blood. This is to say, the book has a body" (2). The embodiment of technologies is a key theme for this work, and I side with Littau here, and also with Hayles: "technologies are embodied because they have their own material specificities as central to understanding how they work as human physiology, psychology, and cognition are to understanding how (human) bodies work" (*Electronic* 112). I want to pursue, and increasingly seriously, this notion that material equipment, such as a codex or e-reader, has a body that is as worthwhile to understand as the body of its human user. We can begin, in the case of reading, by noting that the claim that a printed book has a body matches reported phenomenological experience, though such reports can be harder to track down as the idea is tightly bound into the language that we use to discuss or describe the medium.

One of the original theorists of new media, J. David Bolter, for instance, considers the role of embodiment in the unification of a book's content with its material form: "The paged book became the physical embodiment, the incarnation, of the text it contained. Incarnation is not too strong a metaphor. Through printing, we have come more and more to anthropomorphize books, to regard each book as a little person with a name, a place (in the library), and a bibliographic life of its own" (*Writing Space* 86). There seems to be more than a degree of subjective report here, rather than a historicised statement of fact, but that's not to say that Bolter's is an idiosyncratic point of view. Littau's wider argument in *Theories of Reading*, for instance, rests upon it, and Hayles also notes that "[a]uthors regularly [think] of their books as offspring [and]...the human form converge[s] with book technologies...in such inert metaphors as footnotes, spine, and appendix" (*Writing Machines* 39). And it's here that we can see that implicit connection with ourselves: we often describe books with the same words that we use to describe elements of our own material existence: "Texts assimilate utterance to the human body. They introduce a feeling for 'headings' in accumulations of knowledge: 'chapter' derives from the Latin *caput*, meaning head (as of the human body). Pages have not only 'heads' but also 'feet', for footnotes" (Ong 100). Do resisters of the new technologies of reading flinch at the thought of separating the book from a body that is so often referred to in such human terms? Such ideas certainly introduce bodies into the discussion of the codex in a non-trivial fashion – e-reading enacts a Cartesian split in the rending of the text's "brain" from the book's body, another variation on technology's dividing us from the felt world.

Hayles' reference to books as offspring, Bolter and Ong's anthropomorphism, and Littau's explicit pronouncement of embodiment all speak to our interactions with printed books being an interaction between two distinctive bodies. We'll pick up on this idea frequently, and with greater significance, extending it out to all material engagements with technologies. For now, however, we simply need to note that physicality matters and it is missed when it seems to be removed from an experience.

1.2.2 E-reading and the corruption of the mind

The second appeal to a decline in the affinitive embodied print relationship stems from a belief in the detrimental effects of the seeming disembodiment of the form. Subjective resistance, as we've seen above, can stem from things "not feeling right" in various ways, but commentators keen to challenge the adoption of the new reading media have increasingly turned to scientific (as well as pseudoscientific and scientistic) arguments in order to establish objective harm in a move away from print. My aim in this short section is not to prove or to refute evidence-based arguments for the negative effects of e-reading, but simply to establish that the discourse is real and, more usefully, represents another significant example of the rhetoric of "unnatural" technology.

Put bluntly, e-reading is often portrayed as making users "stupid." This is perhaps the most provocative of the claims that I've taken from the popular discussion, but in variations on its explicit wording it might also be both the most frequently heard and most clearly indicative of our dysfunctional understanding of the role of technology in our lives. The use of the word "stupid" here comes from a popular and much discussed article in *The Atlantic* by Nicholas Carr: "Is Google Making Us Stupid?" Carr's focus in that piece, and in his book *The Shallows*, is on the glut of competing information online, and whether our outsourcing of memory to search engines and internet databases is making us, as individuals, intellectually deficient in one way or another, be that in terms of attention span or brute IQ (the former levelled at readers and researchers of any stripe, the latter at young children still learning how to read).[24]

There is a growing body of competing literature on the cognitive effects of reading on screen, and Mangen's article "Hypertext Fiction Reading: Haptics and Immersion" provides an excellent overview of academic research into the impacts on cognition and phenomenological experience of reading on screen. What is immediately obvious from the paper (published in 2008) and the book which followed,[25] however, is that the field is, as yet, small in breadth and scope, often unquestioningly applies findings based on web surfing or hypertext fiction reading onto linear e-reading on specialised equipment,[26] and fails its own calls for further

---

[24] Carr's argument isn't specifically about e-readers, but the internet and its associated reading modes and practices have become indelibly linked to, and within, the new reading equipment resulting in Carr's work being cited frequently in discussions of a wide variety of text-based screen media (e.g., we'll shortly see *The Shallows* being cited by Hayles).

[25] The 2009 *The Impact of Digital Reading on Immersive Fiction Reading*.

[26] It is telling that, in a talk at the *Unbound Book* conference in May 2011 Mangen was still citing much the same studies, and forced to make the same conflations between hypertext, web surfing, and e-reading. No empirical study that she cites comes after 2007, that is, after the first generation of Kindle, bar one 2011 study (Ackerman and Goldsmith "Metacognitive Regulation of Text Learning: On Screen Versus on Paper") which only deals with desktop browsing. Things have improved recently (see below), but this demonstrates the youth of the field, particularly with regards to the most recent reading equipment. This means that there has been very little in the way of longitudinal studies; at the time of writing the Kindle is only seven years old, the iPad just four – it's impossible to speak about their long-term effects, hence my call for the interpretation of extensive anecdotal report.

phenomenologically sensitive research: "Until quite recently...issues of materiality have been largely neglected in reading research overall. Several studies point to the importance of addressing the multisensory dimension of digital reading...without really pursuing the issue any further" (Mangen, "Hypertext..." 405).[27] The kind of studies Mangen calls for are beginning to emerge (often through Mangen's own work),[28] but their evidence remains inconclusive, in no small part due to the ever-changing diversity of screens that readers can face and the relative newness of this class of artefacts.

As researchers attempt to investigate variations on theories such as Carr's, one fact does become clear: the workings of the brain have entered the discourse on the subject at every level – academic, journalistic, casual blogging – and, often frustratingly, to the detriment of the rest of the body. This said, the neurological theories, and their widespread appeal, are well worth investigating.

Most research of this kind considers the alternative reading modes prompted by internet surfing, but the issues of distraction, hyperlinking, and varied writing spaces are increasingly relevant to tablet computing with the lines between laptop and desktop browsing and e-reader-type reading becoming increasingly blurred. Maryanne Wolf, in her popular work on the neuroscience of reading, *Proust and the Squid*, offers a number of convincing arguments for why we might express concerns with the ways in which written content is consumed online, particularly by young or otherwise still learning readers:

> The basic visual and linguistic processes might be identical, but [will] the more time-demanding, probative, analytical, and creative aspects of comprehension be foreshortened? Or does the potential added

---

[27] Mangen cites the following studies in this regard: Back "The Reading Senses"; Bearne "Rethinking Literacy: Communication, Representation and Text"; Kress *Literacy in the New Media Age*; Mackey *Literacies Across Media: Playing the Text*; Mackey *Mapping Recreational Literacies: Contemporary Adults at Play*; Merchant "Writing the Future in the Digital Age"; Walsh "The 'Textual Shift': Examining the Reading Process With Print, Visual and Multimodal Texts"; Walsh, Asha, and Sprainger "Reading Digital Texts."

[28] See, for example, Mangen's recent "Putting the Body Back into Reading"; "Reading Linear Texts on Paper Versus Computer Screen"; and "Cognitive Implications of New Media." Mangen has also spearheaded a research network of over 25 countries under a COST European networking grant (E-Read: Electronic-Reading in the Age of Digitisation) to further research in this area, particularly around phenomenological and empirical studies.

information from hyperlinked text contribute to the development of children's thinking? Can we preserve the constructive dimension of reading in our children alongside their growing abilities to perform multiple tasks and to integrate ever expanding amounts of information?. (16)

This quotation holds a number of the most pressing concerns for resistance that draws on the altering neurology of reading: the effects of the abundance and availability of information; the intrusion of visual culture into the typographic; the shortening of attention; and last (and made very much least) the possibility that new reading modes might be beneficial, in a supportive capacity, to childhood development. The significance (and our simultaneous ignorance) of these issues shouldn't be underestimated, but as the internet and digital reading practices become increasingly important to the lives of billions of people the discussion becomes increasingly muddied, particularly as conclusions are being drawn from the results of relatively few studies with small sample sizes, unpredictable control groups, and, for obvious reasons, a limited sense of any long-term effects. In contrast to Wolf's work for instance, the BBC reported that internet use was good for the brain, at least for older users attempting to stave off dementia through increased mental activity: "Lead researcher Professor Gary Small said:... 'Internet searching engages complicated brain activity, which may help exercise and improve brain function'" ("Internet use 'good for the brain'").[29] But this kind of argument, based on brain imaging indicating increased activity during web searches, prompted Caleb Crain to write a counter in *The New Yorker* that extra brain activity is probably an argument *against* digitisation:

> the journalist Steven Johnson [writer of *Everything Bad is Good for You*] argued that since we value reading for "exercising the mind" [similar to Small's argument] we should value electronic media for offering a superior "cognitive workout." But, if [Maryanne] Wolf's evidence is right, Johnson's metaphor of exercise is misguided. When reading goes well... it feels effortless... It makes you smarter because it leaves more of your brain alone. ("Twilight of the Books")

---

[29] For more on this effect see the original study, widely reported with commentary from Gary Small et al. "Your Brain on Google..."

Such circles of evidence, typically drawing on reports of reports of primary research, are common; commentators are citing just a few other writers and researchers because the diversity of studies and discussion simply isn't there, let alone a consensus view. And this is another reason for my focus on folk phenomenological reports here; the dataset is bigger, broader, and reveals what phenomenological philosophers and cognitive scientists have long known: the body matters, and no discussion of cognitive effects is complete without considering its role (and this is the subject of the third chapter).

The brain has become a battleground for the new portable reading equipment, but there just isn't the published work which compares reading on screen to reading from the page; no study has put anyone in an fMRI, or any other type of machine, with a printed copy of *A Tale of Two Cities* and a digital edition to see what's different, no one has yet compared like for like. Maybe there's nothing even to be seen about the effects of e-reading in these patterns of oxygenated blood flow. But still we have a great many commentators seeking to reduce the argument to the pernicious impacts of "the screen" on our intelligence rather than any cultural or content issue. Mark Bauerlein's *The Dumbest Generation: How the Digital Age Stupefies Young Americans and Jeopardizes Our Future (or, Don't Trust Anyone Under 30)* perhaps best typifies this reductionist approach: midway into his argument he argues that "[t]he screen...promotes multitasking and discourages single-tasking, hampering the deliberate focus on a single text, a discrete problem" (94). The screen, in and of itself, often becomes the enemy in Bauerlein's hands.[30]

The titles, let alone the content of popular works on the fate of thought for a generation of digital readers and internet users, though not often as blunt as that of Bauerlein (or Carr's first provocation), are still telling: popular books on the impact of digital content and equipment on the

---

[30] It's worth noting that Bauerlein's "evidence" is deeply misguided at times. He cites statistics of slight average reading performance decreases and limited attendance of the opera and art galleries as evidence of a lack of comprehension and engagement whilst ignoring the radical changes in U.S. demographics and financial stability across the period of study. I don't wish to dwell on Bauerlein's book, it can only be held up as a straw man, but for whatever reasons (perhaps simply the provocative title) his study is highly visible in these discussions and some of the language so sufficiently similar to that of the broad popular debate as to be usefully indicative of a more general attitude. For a full discussion of the flaws in one of the most significant studies that Bauerlein cites see Steven Johnson's *Guardian* article "Dawn of the Digital Natives"; for further discussion of some of the specific issues with Bauerlein's analysis see my post at *Teleread*, "Maybe the Dumbest Generation Came Before Us."

brain, cognition, and society include Maggie Jackson *Distracted: The Erosion of Attention and the Coming Dark Age;* Gary Small *iBrain: Surviving the Technological Alteration of the Modern Mind*; Andrew Keen *The Cult of the Amateur*; Susan Jacoby *The Age of American Unreason*; Lee Siegel *Against the Machine*; Larry Rosen *iDisorder: Understanding Our Obsession with Technology and Overcoming Its Hold on Us*; and David L. Ulin *The Lost Art of Reading: Why Books Matter in a Distracted Time.* For all the very real and relevant concerns expressed by work of this kind, when the worst criticisms of the internet and e-reading are frothed into broad generalisations of what the new equipment is doing to its users I can't help but think that the more puritanical commentators are increasingly envisioning people in the vein of William Gibson's "slitscan" viewers, that our screens alone produce a creature

> best visualized as a vicious, lazy, profoundly ignorant, perpetually hungry organism craving the warm god-flesh of the anointed. Personally, I like to imagine something the size of a baby hippo, the color of a week-old boiled potato that lives by itself, in the dark, in a doublewide on the outskirts of Topeka. It's covered with eyes and it sweats constantly. The sweat runs into those eyes and makes them sting. It has no mouth...no genitals, and can only express its mute extremes of murderous rage and infantile desire by changing the channels on a universal remote. (*Idoru* 28–29)

Hayles, in contrast, offers a far more balanced critique of the effects of screen reading, combining her own phenomenological reports with her experience as a university professor training hundreds of students in close reading over years of changing approaches and attitudes to the print reading experience, and also drawing on the research presented by Carr in *The Shallows*. She is most interested in the information readers glean from surfing online, a potentially more distracted mode:

> The small distractions involved with hypertext and web reading – clicking on links, navigating a page, scrolling down or up, and so on – increases the cognitive load on working memory and thereby reduces the amount of new material it can hold. With linear reading, by contrast, the cognitive load is at a minimum, precisely because eye movements are more routine and fewer decisions need to be made about how to read the material and in what order. Hence the transfer to long-term memory happens more efficiently, especially

when readers re-read passages and pause to reflect on them as they go along. (*How We Think* 64)

Here we have a real concern, and one supported by Mangen's research as well as that collected by Carr in his chapter "The Juggler's Brain." There *is* evidence that certain kinds of screen reading can tax readers more than a simpler codex equivalent, but how well this might be adapted to over time, particularly if there are changes in reading pedagogy and the types of artefacts used to access content, remains uncertain.[31]

This full range of responses to cognitive effects, from knee-jerk refusal to considered interpretation, again cluster around the now familiar idea: that reading on screens appears unnatural. It presents a challenge; it increases cognitive load (to such an extent that, in Small's research at least, it represents a kind of workout); it impedes learning; it doesn't feel right; it doesn't feel like reading; etc. But Hayles, again reflective of her far more nuanced approach, offers us a way out alongside her warnings against the impact of the screen:

Hyper reading, which includes skimming, scanning, fragmenting, and juxtaposing texts, is a strategic response to an information-intensive environment, aiming to conserve attention by quickly identifying relevant information, so that only relatively few portions of a given text are actually read. Hyper reading correlates, I suggest, with hyper attention, a cognitive mode that has a low threshold for boredom, alternates flexibly between different information streams, and prefers a high level of stimulation. Close reading, by contrast, correlates with deep attention, the cognitive mode traditionally associated with the humanities that prefers a single information stream, focuses on a single cultural object for a relatively long time, and has a high tolerance for boredom. (*How We Think* 12)

In this comparison of two reading modes – "hyper reading," a response to the sheer wealth of information available, and "close reading," required by the attentive analysis of particular, often complex, texts – Hayles shows us what I will argue is the real "natural" when it comes to our use of all technology: adaptation. Though close reading still comes across as (at least semantically) privileged in Hayles' discussion, there is a space made for new modes to offer practical responses to new environments

---

[31] Again, the longitudinal studies just aren't there, but I'll pick up on the effect of familiarity and expertise from the next chapter and throughout.

of use. The rest of this book will be dedicated to understanding exactly what goes on during such adjustments to technology, and better understanding these phenomena necessarily challenges any sense of the essence of technology as being somehow against our natures.

New technologies, and various forms of e-reading specifically, can be frustrating, challenging, unsettling, and, possibly, detrimental to our health, happiness, or social, emotional, or intellectual intelligence. This is too simple a claim, of course; technologies don't have fixed effects that consistently manifest during predictable use by all users, and this is something that we'll return to in later chapters. But let's take for granted, for the moment, that adopting something new is typically a challenge, whether that be something new in an established field (e.g., e-reading adopted over print reading) or the creation of a new field entirely (e.g., the first automobiles). There are challenges to be faced in learning something new, but, overwhelmingly, what humans do is adapt, shaping themselves alongside the iterations of the artefacts that they surround themselves with. Washing machines, and bicycles, and desktop fans, air conditioning, televisions, stereos, and whole hosts of other technical things now blur into a background of simple, functional stuff in every developed country.[32] Each of these items has its own effects, but they're now not typically frustrating, though each of them had to be adapted to, individually, socially, economically, and culturally.

We can heed the warnings from reports and research into the effects of e-reading, and of all other technologies, without relying on them as evidence for the *inherent nature* of those objects; that is a discussion that we can still always have. We need an account of technology that can deal with how people adapt themselves and the things around them, that can reflect on how we make initially frustrating or challenging artefacts into the most natural things in the world. In miniature such an account is an explanation of why I don't believe that we should fear e-reading; more broadly it represents a move away from a normative idea of what is natural for humans, avoiding a Whig historical conception of linear progress, as well as simplistic utopian or dystopian teleologies of technological adoption. But before I can start to make that case, I need to finish this chapter by establishing a crucial point in two parts: we must unpick the notion of technology as an unnatural force; technology sits at the heart of who we are.

---

[32] This will do for now, but I will go on to argue that every "developing" nation already has its own technical backdrop, a backdrop that at least problematises the notion of development.

1.3 The unnatural codex

I want to challenge the notion of technology being unnatural by continuing the discussion of e-reading before moving the findings back out to broader technological concerns. As we've seen, the idea that new reading technology is unnatural has two distinctive features: that the new artefacts act as an overly complex layer of visceral insulation, and that reading on screen, with its myriad potential distractions, might have negative cognitive effects, roughly a split between body and mind. The other side to these implications, of course, is that printed book reading *doesn't* suffer from such drawbacks and is, in fact, the more natural mode of engagement. Let's pick up the e-reading debate again with another feature that separates the mediums and causes complaint: non-linear reading.

One of the things offered by reading on screen, and particularly online, is the ability to jump rapidly between elements of information through a system of hyperlinks. Clicking on an underlined blue word is the most obvious example of this, still occupying most users' conception of how text functions online. But there is an increasing expectation for searchable words; clickable pictures; swipe-able screens; triggered links; videos with embedded links; etc. – more information is seemingly always to hand. This leads to the criticisms that we've seen: that the previously undistracted reading of print is now more liable to be disrupted, that such linking and mixed media gets in the way of the task at hand and a new non-linear mode of thought will be fostered in its wake. In a largely positive vein we saw Hayles' suggestion of "hyper reading" as a successful adaptation to this new information environment, but Bauerlein, attempting to explain his problems with the increase in time spent online, states that

> the cultivation of nonlinear, nonhierarchical, nonsequential thought patterns through Web reading now transpires on top of a thin and cracking foundation of print reading. For the linear, hierarchical, sequential thinking solicited by books has a shaky hold on the youthful mind, and as teens and young adults read linear texts in a linear fashion less and less, the less they engage in sustained linear thinking. (141)

The argument, Bauerlein assumes, is so clear as to not need stating explicitly: linear reading is so beneficial in and of itself that any move away from it will be detrimental to the equally beneficial activity of linear thinking. Again, the presumption here is that linearity is natural and maybe moral; recall Birkerts' earlier claim, that "[w]hat [codex] reading does, ultimately, is keep alive the dangerous and exhilarating idea that...God or no God, life has a unitary pattern inscribed within it." This "goodness," written into the bodies of the artefacts and the thinking that they provoke, might well seem fundamental, but it can be challenged, and from a variety of quarters – such a contestation, then, has much to say about our attitudes towards newness and to the role of technology in our lives.

In a paper from 1945, "As We May Think," the engineer and early computer scientist Vannevar Bush described a machine for organising information in libraries that was intentionally non-linear, explicitly envisioned as an extension of the mind, and one of the first examples of a planned hypertextual device. Bush called his then hypothetical system the "Memex" (a memory index or extender).[33] Hayles refers to Bush's proposal in her own discussion of linearity in reading, and comes to side, at least partially, with Bauerlein and Birkerts: "Bush [had] argued that...the Memex was superior because it worked the way the mind works, through association. Kaye[34] was not sure the claim was correct. Certainly she sometimes caught herself thinking through association, but logical ordering and linear sequencing were also important" (*Writing Machines* 75). Hayles' language here is revealing: associative thinking is something to be "caught" doing, even if it is important, whereas *logical* ordering and linear sequencing one can be more open about. The term "logical" is weighted, becoming synonymous for "sensible"; the associative thinking prompted by non-linear access online (and in electronic texts that can be interrupted by additional material in the same space) is implicitly illogical, even insensible, certainly for commentators such as Bauerlein and Birkerts, if more problematically for Hayles. And if the codex's linearity matches or supports the best of the ways in which people can cognise, then it is important to preserve, and possibly to the active resistance of the new digital reading technologies. We therefore need to consider whether that enduring faith in the printed book matches up with psychological reality.

The author Hans Magnus Enzensberger offers us a way-in by considering the nature of writing and what it promotes in our thinking:

> The formalization of written language permits and encourages the repression of opposition. In speech, unresolved contradictions betray themselves by pauses, hesitations, slips of the tongue, repetitions, anacoluthons [changes of syntax], quite apart from phrasing, mimicry, gesticulations, pace and volume. The aesthetic of written language scorns such involuntary factors as 'mistakes.' It demands, explicitly or implicitly, the smoothing out of contradictions,

---

[33] A simulated version of the Memex is now available online (*Memex Sim*).
[34] Hayles' fictionalisation of her younger self in *Writing Machines*.

rationalization, regularization of the spoken form irrespective of content (33).

Enzensberger is in line with the commentators that we've already encountered in believing that, regardless or in spite of the content that it presents or fosters, the formal qualities of a medium can shape thought. But he also sets us down an interesting path: if our most beneficial thought processes are the processes most related to our nature then ironically the codex might well be the more unnatural shaper of cognition in its only preserving a smooth linear mode. In much the same way as we might look at a typical countryside image and think that it is pleasingly "natural," forgetting the centuries of human landscaping that have gone into its construction, so many readers have consumed printed books and reported that they appear to model their thoughts accurately. My contention is that perhaps they should have asked if their cognition has instead been modelled to fit the printed page. As Sergio Cicconi puts it:

> Chirographic [hand] writing, and, later, typographic writing, have strongly modelled the organization of our thoughts, so much that now we tend to think of the linear and propositional structures of printed books as the most faithful representations of the way we organize thinking. But in spite of the paradigmatization of the "printed-thought," a printed text is a very vague (and artificial) approximation of the flow of our thoughts.

We think in a "print" way, not because that's our "natural" way to think, but because our society has developed a heuristic of codex reading, standardising the gestures of interaction and establishing reinforcing structures of use such as the privileging of "clean" linear thought over the complexity or realities of association, selecting for its strengths in a very specific way. This has shaped our minds, and also our culture, so that organised linear thought has long been prided as intellectually superior, as a sign of the brain working at its peak.[35] There is no doubt that organising one's

---

[35] The psychologist and philosopher Alison Gopnik offers a great satirisation of the impact of literacy in her contribution to *Is The Internet Changing the Way You Think?*:
Yes, I know reading has given me a powerful new source of information. But is it worth the isolation, or the damage to dialog and memorization that Socrates foresaw? Studies show, in fact, that I've become involuntarily *compelled* to read; I can't keep myself from decoding letters. Reading has even reshaped my brain: Cortical areas that once were devoted to vision and speech have been hijacked by print. Instead of learning through practice and apprenticeship, I've become dependent on lectures and textbooks. And look at the toll of dyslexia and attention disorders and learning disabilities – all signs that our brains

thoughts into a neat and cohesive narrative is often useful or even essential to action, but to suggest that it is our default, or even most productive state is a mistake sustained by the equating of mental efficacy with the inflexible drive forward of the printed word. It may be the case that computer- and e-reader-based reading which *combines* the hypertextual *with* the linear is the form most suited to our natures, or perhaps yet more applicable is some as yet unknown form, but what is certain is that as long as the codex is bound up in a doctrinal naturalness to which all else is automatically inferior then we are far less likely to be allowed to discover it, hence the importance of questioning our pride in linearity.

A printed book's materiality certainly doesn't actively promote non-linear cognitive work, but the multiple reading panes of internet browser tabs, or the parallel processing of a tablet or mobile phone that allows rapid switching between tasks may enact or perform sympathetically with such a way of thinking. If "natural" thought represents a balance struck between active, associative, or metaphorical thinking with reductions down to a single stream for focus as required, then perhaps equipment which is able to accommodate and promote both of these modes should not be the one to be denigrated, ahead of time, as making us "stupid." Linear thought, for all of its importance for close reading and other attentive actions, resembles professional athletic training: a narrow optimisation rather than a strengthening of a fundamental approach to all action. Focus and the linear, reductive, and self-narrativising aspects of complex thought *are* vital, particularly for complex cultural negotiations, but, following interpretation, information devoid of contextual and imaginative linking can end up as simplified rote learning, surely a starting point for more active engagement and association rather than the ultimate cognitive aspiration.

Evidence for our innate propensity towards non-linear thinking can be found throughout Philosophy, the arts, and folk phenomenological report, and each of these might act as a provocation to Cognitive Science as samples of phenomenological experience.[36]

---

were not designed to deal with such a profoundly unnatural technology." ("Incomprehensible Visitors..." 271)

[36] As the cognitive neuroscientist Merlin Donald notes, for instance,
[l]iterature affords us a great luxury, one that we lack completely in the clinical study of consciousness because even the most experienced clinicians remain outsiders to their patients' minds and are constrained by the formal, conventional nature of their encounters with others. Fiction...provides a...reality check, built from expert observations but from the inside. For this reason alone, literature must become part of our database. It is perhaps the most articulate source we have on the phenomenology of human experience. (*So Rare* 78)

For a philosophical perspective on non-linear thought, though one, admittedly, fascinated by Neuroscience, we can turn to Giles Deleuze and Félix Guattari:

> Thought is not arborescent, and the brain is not a rooted or ramified matter. What are wrongly called "dendrites" do not assure the connection of neurons in a continuous fabric. The discontinuity between cells, the role of the axons, the functioning of the synapses, the existence of synaptic microfissures, the leap each message makes across these fissures, make the brain a multiplicity...a whole uncertain, probabilistic system...Many people have a tree growing in their heads but the brain itself is much more grass than tree. (*A Thousand Plateaus* 17)

Deleuze and Guattari see, in the structure of the brain itself, a non-linear and uncertain system at odds with the "arborealised" thoughts that can emerge from it. For them, the privileging of linear and hierarchical thinking is the "unnatural" mode, part of the disorder or pathology of late capitalist society; it is the tree which should be hidden, or escaped, not the multiple relations of the grass which lies at our heart. For Deleuze and Guattari one should avoid being "caught" thinking arboreally.

Similarly, turning to literature, it is in the context of a privileged non-linearity that we can understand the words from the container which holds B.S. Johnson's *The Unfortunates*, an unbound "book in a box" designed to be read in any order: "The book form fails to capture the 'truth to life.'" The materiality of the codex matches neither the simultaneous stories of the world, nor the minds that try to comprehend them, and literature contains a myriad of these small rebellions against the inheritance of linear order. We can see these concerns expressed, for instance, by poetry's attempting to apprehend the phenomenological experience of cognition, poetry being perhaps *the* historical site of the struggle between words and the perceptions of worlds. Jorie Graham, for instance, in the second "Prayer" of her 2003 collection *Never*, deploys a specific device, rounded and squared parentheses, to layer meaning against the linear drive of the line:

> I love the idea of consequence.
>
> Is that itself consequence – (the idea)?
>
> I have known you to be cheap
>
> (as in not willing to pay out the extra
>
> > length of

blessing, weather, ignorance – all other
[you name them] forms of exodus). (14)

There's a tendency in *Never*'s layered voices to use the rounded parentheses for corrections in the poetic voice, and the squared brackets for asides in a cadence that seems to exist outside of the dominant style of the collection; this is a particularly print-able device where the distinction between the two punctuation marks is clear. We can see, in an earlier poem from the same work, how Graham again builds up a discussion with herself in squared brackets midway through a line, detailing how the writing of the poem itself triggered memories which first interrupt, then aid her composition:

skeletal diminuendos of glancings as they
ascend the manifest up towards its upper reaches – soil,
timothy, stone, manyness of stone, non-mortared
                  build-up of it –
mistings of just-above-stone where the
two of them meet, manifest, un-manifest [and how they
                  could not
know who was looking at them][and that I was from
another country][down to the very movement of my lips]
[show me a word I can use][and how all that you say
is taken from you, they take it, just like
that it becomes smoke] smoke rising here as mist off the heavy
                  topmost stones. ("Philosopher's Stone" 7)

Writers have long lamented the inability of words to capture the complexity of the world and of thought, but Graham, here, does not seem unhappy with words per se, only with the impossibility of the ever-driving-onward line to do justice to the way that she is thinking. In this second quotation, the mess of images that seem to contribute to showing her a word that she can use, the progression toward the word "smoke," is inadequately captured, still seemingly let down by linearity, by the fact that there can't be layers at once; the variety of a moment of cognising in time can only be expressed, equated, with forward motion. Parentheses function, here, as a material metaphor; a punctuation tool that has long been used to signal the problem, in speech and in writing, of capturing messy thoughts, tangents, associations, and novel promptings, is put to work: "this isn't how we think, don't forget that, but also don't get over it by pretending that we don't think at all." More simply, Graham doesn't want her lines clean because thinking is messy. In a discussion of Graham's poetry, Helen Vendler argues that

> [t]he appetitiveness of the mind, and the infinity of the world's stimuli, generate the excess of Graham's long horizontal lines, which generate, in their turn, her long vertical sentences. Any given poetic idea begins to produce, in Graham, a version of an aesthetic Big Bang with its vertiginous perceptual expansion and its receding conceptual distances. (54)

It is in this expansion of perception that Vendler sees Graham's interest in the mess, the layers, her chance to "feel her way into the heterogeneity, simultaneity, chromatic change, spontaneity and self-correction present in all acts of extended noticing" (54). The associations given off by the act of considering are natural for Graham and, regardless of the historical strictures of the line, they deserve, they *need* to be presented. There's something worth flagging here, though it won't prove relevant until a later discussion: in her observations and their non-linear expression, Graham depicts cognition as an active process that occurs in tandem with the thinker's being situated in the world. This, I hope, will come to seem our natural way of thinking about and through technology – I certainly intend to leave neat linearities behind from here on out.

Related to Graham's take on associative thinking, we can also look to Walter Ong's discussions of reading technologies and literacy. Ong considers linear writing's failure to represent our innately associative thinking in terms of "redundancy": in one sense needless additional information (i.e., repetition or off topic associations are redundant), but in information theoretical terms, redundancy actually protects messages from corruption, distortion, or misconception.[37] By providing additional contextual information and repetition, redundant speech makes it less important for any one particle of information to be interpreted correctly:

---

[37] See, for instance, James Gleick's *The Information* (21–34) for a discussion of the redundancy built into both spoken English and African talking drum languages. Because every word in the drum language is so similar to every other word, their meanings become hugely context dependent, so much so that the average drum language sentence is eight times longer than its English equivalent (which, as Gleick demonstrates, also has its own built in level of redundancy based around contextual clues), fleshing out the context for each of its elements.

> Since redundancy characterizes oral thought and speech, it is in a profound sense more natural to thought and speech than is sparse linearity. Sparsely linear or analytic thought and speech is an artificial creation, structured by the technology of writing...which imposes some kind of strain on the psyche in preventing expression from falling into its more natural patterns. (Ong 40)

Ong sees in our originary patterns of speech a more natural expression, of necessity, in relation to the patterns of our thinking, one that is redundant, associative, and (implicitly) richer. The artifice of the technology of writing "strains" the psyche and its "natural" processes, and though Ong possibly overstates a problem here (linearity isn't entirely unnatural, just rarer than we might assume, and certainly not our sole means of thinking well), such strain can also be seen underlying Graham's parentheses and Johnson's mutable novel.

Such ideas don't mark an attack on linearity, or on the page, but they do highlight its unnaturalness, offering a riposte to those who would position screens as the unnatural antithesis of the codex's coherence to human thought, to those for whom the new practices prompted by the digital must always be inferior or detrimental. These instances highlight the contortions that thought must go through in order to be represented in a medium that equates forward motion with meaning, with at once time passed and the filling in of detail within a moment.

1.4 Technogenesis

We now have a basis on which we might start to discuss the nature of technology. We've seen a rhetoric of unnaturalness emerge based around technology's interruption of embodied experience and of cognition, but we've also started to challenge a simple reliance on what initially appears to be natural: particularly when it comes to written language, our thought has always-already been prepared by earlier technologies; there are other ways to think.

I'll conclude this chapter with a final challenge to the idea of technology as being something "unnatural," allying my argument with "technogenesis," a term drawn from the work of the philosopher Bernard Stiegler to describe the coming into being of humans *with* technology, and humanity's continued coevolution with our technological artefacts. I will use the term here to refer to this full range: that humans couldn't have been what they are, or continue to become as they do, without the artefacts that they put to use – in reality, and in terms of the shape of this work, this must be both the entry point and conclusion to understanding the technological.

1.4.1 Dismodernism

To enter into this claim, I want to briefly consider the disability theorist Lennard Davis' assertion that humans are always in some way dependent, always requiring support from their world. This might seem a strange turn, but if we can assume Davis' argument then we can more deeply understand the essential role that tools play in the nature of being human and set the ground for a definition of technology (to be pursued in the next chapter) which includes a huge array of equipment that extends our ability to act.

Davis' argument, presented in "The End of Identity Politics," rests on the idea that everyone is born impaired to some extent; there is no unaugmented state of being that we deviate from or can work our way back to; everybody who lives requires support. Davis describes the understanding of this truth as "dismodernist," arguing that though postmodernism did an excellent job in starting to break down notions of essential identity formations it still rests upon a fixed humanist ideal. Dismodernism, then, is a posthuman (i.e., against Humanist) position aimed at breaking down an ignored essentialism. Davis' articulation of this is worth quoting at some length:

> Politics have been directed toward making all identities equal under a model of the rights of the dominant, often white, male, "normal" subject. In a dismodernist mode, the ideal is not a hypostatization of the normal (that is, dominant) subject, but aims to create a new category based on the partial, incomplete subject whose realization is not autonomy and independence but dependency and interdependence. This is a very different notion from subjectivity organized around wounded identities; rather, all humans are seen as wounded. Wounds are not the result of oppression, but rather the other way around. Protections are not inherent, endowed by the creator, but created by society at large and administered to all...The dismodernist subject is in fact disabled, only completed by technology and by interventions...As the quadriplegic is incomplete without the motorized wheelchair and the controls manipulated by the mouth or tongue, so the citizen is incomplete without information technology, protective legislation, and globalized forms of securing order and peace...[T]he by now outdated postmodern subject is a ruse to disguise the hegemony of normalcy. (240–241)

For Davis, a truly posthuman stance would allow us to recognise just how dependent we all already are. Rather than a Humanist faith in a "pure" subject, this stance calls for us to recognise the extent to which we each benefit from existing systems, legitimate our various uses of those systems, and raise our empathy for those who require different uses, or different systems, for their own self-actualisation. Dismodernism would mark the true end of the notion that there is some normal subject from which we might deviate; it challenges "somatonormativity," the idea that there is a complete and unchanging human body that ideally requires no external support in order to be truthfully itself; for Davis this is the last great myth.

If we take even the simplest point from Davis' argument here it supports the technogenetic argument that I want to make: humans require external support simply in order to be. We need the support of our parents, and the social structures that we move into; we need the support of shelter, and tools, and farming or hunting groups; and, to be a part of our societies, we need to take on all of the myriad devices that have become normalised, whether that be knives and instruments and bows and blowpipes, or washing machines, mobile phones, cars, and computers. The discrimination that Davis identifies is discrimination against people who require particular *kinds* of support, a discrimination that presupposes a norm of radical autonomy as it assumes that those who are doing the discriminating function without support. This is, strangely, much the same discourse as that developed by those who would oppose technology in general, and e-reading in particular, on the grounds of unnaturalness: such thinking presupposes an unrealistic normal of unaided life and judges from that position – that a new technology gets in the way of "natural" human interactions with the world suggests that we can experience the world "purely" and unaided in some ideal absence of technology. Similarly, arguments that e-reading negatively affects our cognition often presuppose a norm of cognition which exists without complex technological interventions and one that is conveniently naturally suited to the affordances of the printed page.

1.4.2 Embodied cognition

In *The Artificial Ape*, Taylor argues that "[t]he idea of humans versus technology is wrong...Technology is at least as critical to our identity as our soft tissues" (189); akin to Davis' dismodernist position, Taylor sees every human as requiring external support in order to be. The use of technology in all human societies, from hand tools and basic weapons to industrial machinery and nuclear bombs, is ubiquitous. If "technology" refers simply to external tools for getting work done, a point that we will interrogate in the next chapter, then we can be certain that humans have technology at their hearts. We can look at the lives led by any nomadic or settled people, in any human habitat, from Inuit *tupiq* to Bedouin *bayt char*, from favelas to penthouses, and the defining feature of homo sapiens' existence is the use of equipment which extends our ability to ensure our thriving survival.

Karl Marx saw this placing of ourselves out into the world through our made artefacts as a fundamental need: "Humanity needs objects as objects of its life-expression...The true human life becomes the externalized life" ("Economic and Philosophical Manuscripts..." 167). As David Rothenberg notes, "[t]echnology is the first act of humanity for Marx. Living through *techne*, we transform nature into history and thus into meaning. We cannot do this just by thinking about our place in the world, but only through working with the world...carving a world out of nature. Because we are human we look for our world outside of ourselves" (74). Recent archaeological evidence offers support for Marx's belief, suggesting that basic tool use, what many would see as the first technological interactions, may have been a part of our hominid ancestry for over three million years[38], but without doubt it has been a part of *Homo sapiens* life since its very beginnings, shaping our social structures, eating practices, and basic survivability. Taylor states this baldly: "There are seven species of great ape on the planet. Six of them live in nature. One cannot live without artificial aid. Humans would die without tools, clothes, fire, and shelter" (*Artificial* 1). Taylor's work demonstrates the intimate and inseparable relationship between humans and their tools evidenced throughout the archaeological record, and suggests that this combination may have been one of the principle factors that allowed our species to evolve. Perhaps his most significant insight is that the creation of carrying slings by our hominid ancestors may have been the turning point for the evolution of both bipedalism and modern human cognitive capacity:

> an inspired female pick[s] up a twisted loop of animal skin – perhaps some sun-toughened membrane at a scavenging site, or some scorched but not burnt pelt from the embers of a bush fire.

---

[38] See Shannon et al. "Evidence for Stone-Tool-Assisted Consumption of Animal Tissues Before 3.39 Million Years Ago at Dikika, Ethiopia."

With the infant seated and the strain off the arm, energy requirements for moving with the child plunge, by a massive average 16 percent. So the pressure to make this discovery, even to remake it more than once, is huge... Although carrying slings do not in themselves drive brain-size increase, they certainly encourage it. Rather than having to fit a larger and larger cranium through a pelvic girdle that has contorted itself to support an upright frame, helpless babies can be catered for in a pouch... And that, of course, is the solution to growing larger brains: you do it once outside the womb. (123–124)

Taylor's thesis starts to demonstrate just how fundamental tool use is to humans: with the theory outlined above, he positions technology so deeply at our heart that we could not have come to exist without it. And it is this position that I would like to take up here: there is no such thing as the human without technology and this is a direct function of our cognition and embodiment being intertwined with and enriched by our use of things out in the world. Technology, I will argue, is frequently a part of our natural cognition and of our embodied experience of the world, and the same branches of Philosophy and Psychology that restitch the Cartesian divide also give us a window onto such technological use. Contemporary Cognitive Science, in particular, expands the study of cognition beyond a simple "brain-bound" model, demonstrating the vital roles that embodiment and environmental support play in the cognitive strategies that we can deploy. For now I'll call this simply "embodied cognition" (and the study of such "embodied Cognitive Science"), but as we progress through this work I'll pull out some of its more distinctive branches. Such work might be seen as falling under a dismodernist ethic by breaking down, as we'll see, the reified boundaries of who or what cognises – its claim is clear: there is always help from outside.

Embodied Cognitive Science is the study of, and belief in the body's capacity to affect, to deeply structure, the activities of the mind, radically opposed to the Cartesian dualism of the Western philosophical tradition. Though not solely a scientific discipline or subset of Psychology, embodied cognitive science often relies on experimental Cognitive Neuroscience and empirical psychological evidence for its claims. Those claims, however, can be both inspired by and in pursuit of philosophical or sociological ends; work in phenomenology, for example, underpinned its inception and can still be found explicitly highlighted in work from, for example, George Lakoff and Mark Johnson,[39] Shaun Gallagher, and Andy Clark.[40]

Gallagher, in *How the Body Shapes the Mind*, makes it clear from the outset of his argument just how fundamental the body in motion is to thought:[41] "In the beginning, that is, at the time of our birth, our human capacities for perception and behavior have already been shaped by our movement" (1). From before we enter the world our bodies have been rehearsing actions which have primed us for existing in our environment, the ways that we'll move, and the ways that we'll take on board information. And from our very first moments in the world movement is established as reciprocal, primed to be responsive to its surrounds: "precisely and quite literally, we can see our own possibilities in the faces of others. The infant, minutes after birth, is capable of imitating the gesture that it sees on the face of another person. It is thus capable of a certain kind of movement that foreshadows intentional action, and that propels it into a human world" (1). For Gallagher, as for most, if not all supporters of embodied cognition, it is the specifics of our bodies in motion, or our desire for motion, that forms a large part of what makes us human.

In this light, Frank Wilson discusses issues that we might readily describe as being the subject of embodied cognition in *The Hand*. Wilson's research takes the idea that our bodies shape all aspects of our mental life and focuses in on the central importance of our hands: "I would," he argues, "[say] that any theory of human intelligence which ignores the interdependence of hand and brain function, the historic origins of that relationship, or the impact of that history on the developmental dynamics in modern humans, is grossly misleading and sterile" (7).[42] Part of what I'm hoping to establish here is a theory of at least an aspect of human intelligence, and as such I want to be sensitive to Wilson's admonishment and to consider the importance of the effectors which directly engage with the majority of our most significant artefacts. It is

---

[39] See *Philosophy in the Flesh*.

[40] See *Natural Born Cyborgs* and *Supersizing the Mind*. For more on the embodied cognition project see Margaret Wilson "Six Views of Embodied Cognition"; the excellent overview of "Embodied Cognition" at the Stanford Encyclopaedia of Philosophy; Lawrence Shapiro *Embodied Cognition*; Francisco J. Varela *The Embodied Mind: Cognitive Science and Human Experience*; and Anthony Chemero *Radical Embodied Cognitive Science*.

[41] Gallagher's project in this work is to develop a language which cuts across phenomenology, Cognitive Science, and experimental Neuropsychology, allowing for a coherent way of discussing the body's influence on cognition.

[42] For further discussion of the role of hands in cognition see Zdravko Radman's collection *The Hand, an Organ of the Mind: What the Manual Tells the Mental*.

no coincidence that it is often our hands and sense of touch that will receive most of the attention here.

Bringing the discussion of embodiment explicitly back to our interest in technology, a number of anthropologists and evolutionary psychologists have argued for the impact of tool use, of the dextrous use of the hands with an object, as being integral to both the development of brain size in our primate ancestors[43] and to using that increased brain power to later create language, language which would, for some cultures, eventually be made concrete and preserved in codices. Drawing on such research, Wilson argues that

> [i]f language and the employment of the hands for tool manufacture and tool use co-evolved – effectively forging a new domain of hominid brain operations and mental potentials that we collectively refer to as "human cognition" – then we *should* find analogous links, or reinforcing effects, between purposive hand use, language, and cognition in the individual histories of living people. (*The Hand* 34)

He goes on to discuss the work of Patricia Greenfield, particularly her study "Language, Tools and Brain: The Ontogeny and Phylogeny of Hierarchically Organized Sequential Behaviour," and its evidence for such links: "These two specific skills (manipulating objects and manipulating words), and the developmental chronology associated with the child's mastery of those skills, proceed in such transparently parallel fashion that the brain must be: (a) applying the same logic or procedural rules to both; and (b) using the same anatomic structures as it does so" (*The Hand* 165). Wilson sees in Greenfield's work clear support for his understanding of an evolved connection between language and the use of the hands in tool use, and further evidence for these assertions can be found in Stanley H. Ambrose's "Paleolithic Technology and Human Evolution," where Ambrose outlines the crafting of early tools as synching with the cognitive and linguistic abilities of early hominids. Striking stones together to produce blades "involve predominantly repetitive coarse motor control (percussion flaking). Primate vocalizations are also repetitive sequences of coarse motor actions." But tools which combine elements

> are hierarchical and involve nonrepetitive fine hand motor control to fit components to each other. Assembling techno-units in different

---

[43] See for instance Stanley H. Ambrose "Paleolithic Technology and Human Evolution"; Beth Preston "Cognition and Tool Use"; and Kathleen Gibson and Tim Ingold *Tools, Language, and Cognition in Human Evolution*.

configurations produces functionally different tools. This is formally analogous to grammatical language, because hierarchical assemblies of sounds produce meaningful phrases and sentences, and changing word order changes meaning. Speech and composite tool manufacture involve sequences of nonrepetitive fine motor control and both are controlled by adjacent areas of the inferior left frontal lobe.[44] (1751–1752)

Ambrose uses this evidence, alongside the broader archaeological record, to support a hypothesis of at least the coexistence and likely the coevolution of language and tool use, each activity reinforcing and manipulating the other reciprocally. Such connections would now seem to come at least somewhat hard-wired with the neural pathways that evolved for effective tool use piggybacked upon by the development of an inbuilt capacity and drive to acquire a symbolic language.

This is unsurprising when we consider the kind of intelligence that already needs to be in place for sustained dextrous tool use with multipart equipment. Reliable tool manufacture requires the knowledge that if, for example, a stick is attached to a stone then it can, as a unit, be a more effective and accurate way of striking a target; a mental image must exist to enable the intentional repeated creation of such objects. The potential for imagination to prefigure or stand-in for some object in the world is as much a precondition for creating tools as it is for developing a sign system. This is why the use of multi-part tools is incredibly rare, close to non-existent in other primates: it's not that they are incapable of using them, or even, in theory, of creating them; the insurmountable challenge lies, for the most part, in their inability to *conceive* of them.

Contemporary research into gesture also appears to support Wilson's assertion that we should find links "between purposive hand use, language, and cognition in the individual histories of living people." Researchers have become increasingly interested in the connection between the body and thought (in no small part due to the debates prompted by embodied cognition), and recent work on gesture sits squarely within the discussion. A 2009 study conducted by Susan Goldin-Meadow, Susan Wagner-Cook, and Zachary Mitchell looked at young students being taught the mathematical concept of "grouping." Grouping is used in the solution of problems where a single term must stand-in for several, for example, $3+2+8 = \underline{} +8$. The students taking part in the study had

---

[44] Ambrose cites Greenfield "Language, Tools and Brain" and Kempler "Disorders of Language and Tool Use" in this regard.

to learn to resolve such equations by finding the single digit which is equivalent to 3+2 to fill in the underlined gap, that is, they must understand the concept of combining numbers to produce an analogue which balances the sum. In order to teach this, tutors were getting students to draw a little "v" shape with their finger under the 3 and the 2, physically tying them together. "Previous research has shown that students who are asked to gesture while talking about math problems are better at learning how to do them. This is true whether the students are told what gestures to make, or whether the gestures are spontaneous" (Campana). Sure enough, students understood the concept significantly faster with the grouping strategy than when the technique was not deployed. But the researchers also found, over the course of the study, that it didn't matter where the students drew the "v" at all, that is, it wasn't necessary to link the 3+2 specifically; it was simply the act of making the gesture which introduced and sublimated the concept in "the student, through the body itself" (Campana).

The findings of this research seem to offer a way-in to understanding some of the (scientifically and philosophically) naïve reports of resistance to digital reading technologies that we encountered earlier in this chapter. Goldwin-Meadow et al. show just how important engaging the body is to learning: the students in the study who made no gestures learned how to cope with the task at a significantly reduced rate in comparison to those who made an arbitrarily placed gesture that nonetheless evoked the grouping concept under discussion. This has implications for understanding a shift in reading practices as the gestures of interacting with codices – the turning of pages, the feel of the remaining and consumed leaves held in each hand, dog-earing (or not), breaking spines (or not), all the physical attributes that make the act what it is – these are all undoubtedly changed when the equipment changes. It may well be therefore, as Birkerts suggests, that there is something in the touching of a physical book that becomes combined with the reception of written knowledge to the extent that the folk phenomenological reports are indicating a genuine drop in capacity as some cognitive aspects of the experience become impoverished. Perhaps the gestures of reading print have come to prime the concept of knowledge acquisition, a particular way of thinking, and, as with the students in the above study, the denial of such gestures may interrupt or delay such thought. Suddenly the question of "is this reading at all?" starts to seem less obscure; in a very real way the act of reading changes, and may become a different way of thinking prompted by, or in the absence of prompting, from the movements of our hands. This starts us down the path of seeing technological interactions as an entangled meeting of mind, body, and tool, each with a changeable part to play.

In an earlier work from 2003, *Hearing Gesture: How Our Hands Help Us Think*, Goldin-Meadow had investigated gesture and found that it is likely to have a function beyond being simply expressive as

- We do it when talking on the phone.
- We do it when talking to ourselves.
- We do it in the dark when no one can see.
- Gesturing increases with task difficulty.
- Gesturing increases when speakers must choose between options.
- Congenitally blind people gesture despite never having seen gestures made and also gesture to interlocutors they also know to be blind.[45]

Gesture, then, is fundamental, like swallowing, breathing, or reaching, something we perform automatically, and yet something which can intimately affect our learning. In *Supersizing the Mind*, Clark discusses the use of gesture as a way of extending cognition out from the brain and into the environment by drawing on Goldin-Meadow's work to show another way, beyond physically enacting a concept as with grouping, in which gesture might explicitly benefit learning. The researchers found that negative gestures would often occur during positive speech, and vice versa, leading Clark to argue that

> [t]he physical act of gesturing...plays an active (not merely expressive) role in learning, reasoning, and cognitive change by providing an alternative (analog, motoric, visuospatial) representational format...Encodings in that special visuomotor format enter...into a kind of ongoing coupled dialectic with encodings in the other verbal format...This...creates points of instability (conflict) whose attempted resolutions [often] move forward our thinking. (125)

In this way, gesture actually appears to take on some of the role of an interlocutor, able to enact or code contrary positions in a temporary physical working memory and produce productive conflict.

What is important to convey here is that when we purposefully, consciously or unconsciously, move our hands in space it plays a demonstrable role in meaning and producing meaning. But in truth we simply

---

[45] These findings are further discussed in Clark's *Supersizing the Mind* (123–124).

don't know the extent to which performing actions that require specific repeated gestures and sequences of actions with relatively uniform equipment might standardise those actions, their place in cognition, and our perception of their effects. Folk phenomenological report might again be seen as a productive source of information prior to, and prompting, further research, and the reports outlined at the start of this chapter certainly seem to point toward a collusion between the actions prompted by print and satisfactory cognitive experience, as if the change in media is producing some thus far inarticulable drop in performance.

1.4.2.1 *Hands On Reading*

A final study deals explicitly with this area, outlining the implications for how our hands might affect our reading. As yet, the research presented has been little discussed outside of Psychology, but it would appear to support the folk phenomenological claim that a change in equipment, and therefore the gestures related to the activity, can have profound effects on the reception of written information. As a relatively little known study with a direct connection to the argument of this chapter, it is worth discussing in some detail.

Christopher Davoli et al.'s 2009 study "When Meaning Matters, Look but Don't Touch" considers the hands' effects on perception, specifically the placement of the hands within the visual field during reading. The paper begins with a review of earlier works which addressed the brain's apprehension of the space surrounding the body (peripersonal space) and in particular around the hands as "several results have suggested that visual processing may be biased toward [this] space";[46] such work "support[s] the·conclusion that vision of the space around the hands is special" (556).

Davoli et al. also note that "[s]everal studies have shown that the actions we perform with our hands can influence how we see[47]...These studies have revealed an intimate relationship between perception and action – in particular, the capacity for the latter to affect the former" (555), and it is into this discussion that the paper is situated. The team wanted to explore how use of the hands affected reading ability, in particular whether reading from a desktop mounted screen, where the hands are kept away from what is to be read, is detrimental to understanding

---

[46] See for example Reed et al. "Hands Up: Attentional Prioritization of Space Near the Hand" and Schendel and Robertson "Reaching Out to See..."

[47] For example Bekkering and Neggers. "Visual Search Is Modulated by Action Intentions" and Fagioli et al. "Intentional Control of Attention..."

semantic content when compared to reading with the hands holding the material. Prior work on the impact of handed action on perception

> suggest[s] that spatial processing is enhanced near the hands... Certainly, reading is a process that requires spatial processing. In particular, efficient reading requires the precise control of movements of attention and the eyes through the text, as well as spatial memory to help retain one's place on the page. It is thus quite possible that the preference for holding one's reading material may be attributable in part to the enhancement of spatial processing that occurs near the hands. (556)

This makes intuitive sense when we consider tool use, an activity which predates any act of reading: we need incredibly precise information about our hands' position in space when we dextrously and accurately manipulate equipment. But reading is, of course, not just about space and arrangement, it also includes content which requires semantic processing in order to produce meaning, and at the time of writing it simply isn't known to what extent the hands affect semantic processing. The study from Davoli et al. marks the "the first test of that question" (Davoli et al. 556).

Davoli's team conducted "Stroop tests," a standard measure of semantic comprehension in experimental Psychology, where colour words (red, green, etc.) are displayed on a computer terminal with the letters either matching (e.g., "red" written in red letters – congruent) or not matching (e.g., "green" written in red letters – incongruent) the content that they spell out. The test is well established in Psychology research, with predictable results;[48] it was only the ways in which participants were required to indicate congruency that led to the surprising outcome of the experiments. Participants completed each round of the task by pressing a button attached to either the left or right hand side of the screen to indicate, for example, congruence and incongruence (i.e., the hands would be in the visual field near to the content to be interpreted), or by pressing buttons to the same effect held on their left or right leg (i.e., the hands would be outside of the visual field and away from the content).

The team saw three potential outcomes for the results: (i) semantic processing, as with spatial processing, would be boosted near the hands; (ii) increased spatial processing comes at the expense of semantic

---

[48] The "Stroop effect," for example, has become the term for the delay in reaction time that occurs in reporting incongruent word/letter colour images. See the original study by J.R. Stroop "Studies of Interference in Serial Verbal Reactions."

processing; (iii) spatial processing is improved, but semantic processing is unaffected (556). The placement of the hands certainly did have an effect on semantic apprehension, but not in the way that predictions extrapolated from our folk phenomenological evidence might have suggested: "the present results...suggest that semantic processing is *impoverished* near the hands. This occurred despite the known spatial-processing enhancements that have been reported" (558, my emphasis).

The team found a statistically significant, and in some cases dramatic, drop in response times when a report of congruence or incongruence was indicated with the hands by the sides of the screen rather than their being indicated with button pushes on the legs. This is striking (and unpredicted by the prior literature): despite the copious reports of preference for reading with a book held in the hands over reading from a desktop screen, Davoli et al.'s results suggest that this might actually be the less effective way of processing semantic information. These results were replicated in two further tests set by the team.

During the interpretation of the results, the paper suggests that the effect demonstrated might represent

> a trade-off between semantic processing and spatial processing that can be altered by hand proximity: The enhanced spatial processing that has been observed near the hands...might be achieved at the expense of semantic processing...Indeed, it seems plausible that visual processing near the hands would be biased toward the spatial properties of objects and away from semantic ones. Objects near the hands may be critically important because they might be objects that need to be grasped or obstacles that should be avoided. (560)

The incredible precision afforded to us by increasing the spatial awareness surrounding our hands – and it really is remarkable, particularly when you compare even the infants of our species to our closest primate relatives – this precision seems to come at a cost, and Davoli et al. identify that cost as manifesting in a decreased semantic understanding. When our hands are near something the drain of producing a heightened spatial awareness seems to interfere with the comprehension of a type of information required by another realm.

How then does this data relate to the origin of the folk phenomenological accounts of a preference for holding written material in the hands? It may be that the spatial component of reading vastly exceeds the importance of semantic processing, but this doesn't seem right; however important orienting yourself within the confines of the page or the sentence might be, surely comprehending what passes before your eyes is at least as, if not even more, significant to the overall experience? But if the reading experience might, in some cases, be improved by removing our hands from visual activity then why, as Davoli's subjects supported in pre-trial interviews, do so many people prefer to consistently print digitised reading material out and settle down with their hands primed to interfere with meaning? Davoli et al. suggest that "it is possible that many prefer to hold a hard copy rather than read on a computer monitor because of expertise in reading in this manner. Might the practice that some have with reading in a certain medium outweigh the potential effects of hand proximity? The resolution of all these issues will require further study" (561). It may be that, though the task of handheld reading is cognitively costly, for expert readers the repetition of particular gestures overcomes or compensates for what might, ironically, be described as an "unnatural" (as in a poor match between the task and our nature) engagement. Somehow we have gotten used to reading in this way; somehow it has come to offer something more significant than the additional semantic processing available through keeping the material at a distance. Familiarity, practice, and experience are the features that stand out as syncing the activity with effective cognition, as opposed to some "natural" fit; expertise can overcome the deficit in the impoverished match between certain equipment use and thought, and so much so that the detrimental (but rehearsed) is experienced as the preferable engagement. Expertise is the thing. Expertise is what's natural to humans. And expertise is what we have to focus on in order to better understand our encounters with technologies of all kinds.

1.5 Coda

This chapter has tried to demonstrate a significant thread of resistance to objects that we might identify as being technological. Such resistance can most often be characterised as identifying technology as something unnatural, with such unnaturalness manifesting in the sensation of being separated from embodied interaction with the world or a corruption of cognitive function. Using e-reading as a case study, I have identified some of the ways in which these trends in the discourse around technology have recently been played out again, and have also begun to critique the idea of e-readers as being somehow unnatural in comparison to the codex by demonstrating the ways in which e-reading more closely coheres, and codex reading runs counter, to ways of thinking that we might identify as natural.

We must therefore come to some preliminary conclusion as to what "natural" even means, at least for our purposes here. To say simply that it is the "not-human" is to construct nature artificially, by sterile difference, and it is all the more unsatisfactory considering that there seems to be an often-deployed understanding that allows for some human interactions as long as they are perceived as visceral. Tribal life or antiquated methods of production are often seen as more "natural" because they are somehow further "in touch" with that non-human world, because they don't seem to have a strongly mediating layer of equipment between the person and things as they are. A trowel is different from an industrial digger in this regard; the former quite clearly allows us to get our hands dirty. The use of the body as far as possible without additional equipment is positioned as the more natural mode of engagement so that when it is said that technology, as a class of objects, is unnatural it is implied that it moves the body further away from the work to be done, or the substance that the work is performed upon (this will be dramatically problematised in the fourth chapter). Inherent to this discussion, then, are two broad kinds of "natural": (i) a pre-human world with regard to which we are most "natural" when we are most inert and (ii) a pre-extension "human nature" which we deviate from when we place a greater distance between our bodies and the work in the world. I will say it plainly: (i) is an ethical/ecological question which should have no bearing on any definition of technology. Doing damage to a pre-human nature (if any such thing exists in your surroundings) with a tool or any other artefact is just a more efficient way of marauding unaided, pulling each blade of grass up with the fingers and kicking over every tree. How we choose to act in the world may well be against a metaphysical or even divine pre-human nature, but it doesn't help us to define the tools with which we act even if those tools seem to promote new abilities. A kitchen knife offers both chopped carrots and a back-alley stabbing, and whilst the availability of these extended abilities are part of what the tool is to us, and therefore a part of defining whether it is a technology or not, which action we choose (or are even provoked) to undertake is not.[49] (i) is part of a discussion of something we might call nature, and may or may not be true, but for our purposes it says nothing about the naturalness or unnaturalness of technology, nor the special cases of codex/e-reading. What technology is has nothing to do with unduly affecting the world; ecological thought is about uses, not classes, of objects.

---

[49] As we'll see in Chapter 3, technology can provoke or nudge, but it doesn't determine actions taken.

(ii) is more relevant for our discussion here, and I hope that we have dispensed with it: the idea that there was some pure pre-technology human state that we keep drifting away from is a fiction, as supported by Taylor's archaeological argument and Davis' dismodernism, a challenge to our prideful individualism. We are, as we will continue to see, at our most human when we are enmeshed in webs of support structures including tools, communities, philosophies, and the environments that we receive and shape.

Print, and the kind of thinking that it is allied with, isn't natural, but neither is e-reading. We can, and should, rephrase this: reading and writing aren't in our nature. This seems to me the only phrasing that uses that word "nature" sensibly: to refer to our innate attributes, rather than to some Romantic ideal. And whilst the particularities of reading and writing are demonstrably unnatural in this regard, seen in the fact that we have to teach our children the skills from scratch in each generation, the use of tools, of technologies, absolutely is natural in this sense for human beings who continually spread at least some of their cognitive processes out from their brains, onto their bodies, and out into the world. As Ong describes it,

> artificiality is natural to human beings... As musicologists well know, it is pointless to object to electronic compositions... on the grounds that the sounds come out of a mechanical contrivance. What do you think the sounds of an organ come out of? Or the sounds of a violin or even of a whistle? The fact is that by using a mechanical contrivance, a violinist or an organist can express something poignantly human that cannot be expressed without the mechanical contrivance... Such shaping of a tool to oneself, learning a technological skill, is hardly dehumanizing. (83)

When we can accept that it is the use of and adaptation to technology that forms an aspect of our nature, rather than a favouring of any particular type of artefacts, then we can come to see why print can be so successful without neatly matching every aspect of our cognition, but also how, though e-reading may seem a challenge to some, it also represents a regime that many users might readily adapt to with practice.

So where does this leave us with regard to technology? With worries about certain artefacts that may well be justified and a repeated discourse of resistance to a class of objects that we call "technology" that is seen as unnatural even as items from that very same class enable us to even be at all. Technology is demonstrably a part of who we are, and it always has been, as a fact of historical record; this is something that we need to fully take on board before we begin to consider the (often neglected) nature of its effects. It also seems that there may be something wrong with our definition of the word, and it's to this problem that we must turn in the next chapter.

2
Beyond Common Sense: Technology by Definition

I ended Chapter 1 by arguing that technology rests at the heart of our existence as human beings – there is no being human without technology. This is an extreme claim, but by better understanding technology and its effects it appears to be true; in the same way that we cannot understand domesticated dogs without human contact, we cannot understand humans independent of technological effects. In Chapter 3, I want to focus more intently on the nature of those effects, but first comes a question, one that always shows up late to the discussion: what is technology?

As I intend to demonstrate, there is no such thing as "technology" by any persuasive definition, nothing we can point to or touch, or describe consistent properties of, particularly in its day-to-day use. This is a problem. It is a remarkably loose term; we might often agree on the objects under discussion – computer: yes, coriander: no – but the specifics of why this might be so are vague. "Technology" is, at best, a consensus description of equipment and practices, but one that is singly unidentified as such. When we group items under the term "synthetic" they have a common property to which we can point: the term describes items that do not occur without processing of some kind. When we identify items, rituals, events, or people as "religious," or religiously significant, there may be no materially quantifiable property, but we can certainly describe why we might consider something or someone as such with little effort, appealing only to their immediate relation to some well-defined cultural construction. "Art" is a perennially challenging category of artefacts and effort, but we can still make appeals to creativity, beauty, craft – there may be extensive arguments and recalibrations, particularly at the margins of what has traditionally been accepted as such, but at least the existence of the discussion makes the fluidity of the categorisation apparent. The same is not true, however, of "technology"; technology somehow just is, remaining resistant to nuanced definition, particularly in common usage. Is a pencil a technology? Is a computer? Are they of the same order of things, or is complexity enough to distinguish between the two? This ambiguity has led to the strange scenario that we have already encountered, where a commentator such as Birkerts can describe reading a book as the most "natural" thing in the world, but reading on a screen as a technological twist too far. This is possible, and acceptable, for the significant reason that most of us do not have a working definition of technology that excludes such an assertion.

What makes a hammer of the same order of objects as an industrial press? Does everyone experience this mysterious parity in identical ways, allowing for the consensus, or does "technology" messily draw together items across a range of unrecognised and un-/under-theorised responses? What sort of impact on our lives might be common to objects defined as technologies? In his study of such foundational questions in the philosophy of technology, Joseph C. Pitt notes that "[i]t would help matters if there was a generally accepted definition of 'technology.' Many definitions exist, but there is little agreement on which one is the best. I propose, therefore, to start from scratch" (1). I want to follow Pitt's lead here, returning to our underlying assumptions about what technology is and does in order to suggest that a viable working definition can and should exclude any contention of unnaturalness, and that any definition that fails to do so doesn't recognise the power that technology wields over every human life. In this chapter I will therefore consider some of the range of prior definitions before attempting my own, one that is sensitive to the history of the term's common and technical use, but also to the nature of technology as fully entangled with human life. I will partially be building on Pitt's own work in this area, his own definition being, in essence, that technology is "humanity at work," with the dramatic inclusivity that that broad idea promotes. But I will also attempt to be more rigorous in the criteria by which we might classify something as being technological, and, crucially, what we might also wish to exclude. This will produce some initially strange, but hopefully illuminating, results; the definition that I want to deploy will see, at least potentially, a book, a hunting dog, and a dancer's body as technologies, but banish the Large Hadron Collider, an ATM machine, and a microwave from the same category.

Through this definition, I want to set up an understanding of technological effects, both on the user and on the artefacts themselves, that we'll cover in the ensuing chapters. But I also want to argue that to better understand the term "technology" is, again, to better understand the specific case of the books that we read, in all of their forms. By more fully exploring the terminology deployed, and its implications, we come closer to understanding the resistance to the new digital reading devices in the context of an "adversary culture" (Marx, "Technology: The Emergence...") with a long history and coherent motivations often aimed, as we have already seen, at the notion of "technology" at large.

In Chapter 1, I sketched a history of resistance to technology in a variety of cultural discourses including Jones' identification of the nineteenth century Luddite's political cause being appropriated in so called "Neo-Luddite" attitudes. The existence of these rich precursors to contemporary resistance should make it unsurprising that a seam of such voices has passed into mainstream consensus. I have already stated that I do not believe that the questioning of technological change or progression should be written off, but saying that there is a valuable reason for questioning or resisting technology suggests that there is an identifiable category of thing that is being resisted, and yet this doesn't seem to be the case. The fluid use that the word "Luddite" has acquired – able to describe both the original mill workers and their campaign against job loss for the sake of machinic efficiency and the Neo-Luddite who elects not to use email because he or she still likes to send paper letters – itself indicates that perhaps something may be amiss. The problem is not solely that the politics has been evacuated from the term, rather that "Luddite" has come to mean any resister of technology, and that this is a category that somehow includes mill equipment and email and seemingly everything in between. Where is the boundary line that delineates which items should be included under the term?

The economist W. Brian Arthur, in his attempt to outline *The Nature of Technology*, similarly agrees with Pitt that "we have no agreement on what the word 'technology' means, no overall theory of how technologies comes into being...and no theory of evolution for technology" (12). As we will see in Chapter 5, there have, in fact, been a great many theories as to how technology comes to pass and grow, but the significant point still stands: there is a dramatic lack of consensus in all aspects of technology; what is being argued about is rarely agreed upon. Pitt, again, echoes this: "Central to my concerns is the disturbing tendency of...social critics and others to speak about 'Technology' as if it were one thing. Try as I may, I cannot find the one thing. I can find automobiles, power stations, even specific government offices, but nowhere can I locate *Technology* pure and simple" (Pitt x). I want to start here by looking at how, historically, we have conceived of this one thing, and to ask "what has gone wrong?"

2.1 Existing definitions and issues

For the computer scientist Alan Kay "[t]echnology is anything that was invented after you were born"; for the engineer and inventor Danny Hillis "[t]echnology is something that doesn't quite work yet" (qtd in Kelly 235 & 237). Both of these, admittedly ironical, descriptions point to the notion of technology as something new, something novel. In a similar vein, Jones notes that "[t]oday, for symbolic purposes, computers are technology" (35). There is also an emergence in these remarks that technology is defined by stuff, physical items, manmade, but this certainly isn't always the case. Emmanuel Mesthene saw technology as "the organization of knowledge for the achievement of practical purposes" (25; qtd in Pitt 10), introducing an immaterial component into his philosophy of the subject. Kelly extends this even further: "technology [is] a force – a vital spirit that throws us forward or pushes against us. Not a thing but a verb" (41). When another philosopher of technology, Langdon Winner, suggested that technology could be defined as having moved from being "something relatively precise, limited, and unimportant to something vague, expansive and highly significant" (8), he came unfortunately close to a definitive description of contemporary usage.

For the Luddites dismantling the machines that threatened to supplant them, the word "technology" was largely unavailable:

> [i]t was seldom used before 1880. Indeed, the founding of the Massachusetts Institute of Technology seems to have been a landmark...in its history; however the *Oxford English Dictionary* cites R.F. Burton's use of "technology" in 1859 to refer to the "practical arts collectively" as the earliest English instance of the inclusive modern usage (...instead of being used to refer to a written work...about the practical arts, "technology" now was used to refer directly to...the actual practice and practitioners). (Marx, "Postmodern Pessimism" 247)

Leo Marx's foundational tracing of the term's modern use[1] is revealing, uniting physical artefacts with an aesthetic and knowing praxis from the outset:

> When the Enlightenment project was being formulated, after 1750... [f]or another century, more or less, the artifacts, the knowledge, and

---

[1] For more on the roots of the modern use of "technology" see Leo Marx "Postmodern Pessimism" (247–249) and his "Technology: The Emergence of a Hazardous Concept." Carl Mitcham's *Thinking Through Technology* also deals with the problems and numerous historical definitions of technology (143–154). See also Robert C. Scharff and Val Dusek *Philosophy of Technology* 206–244.

the practices later to be embraced by "technology" would continue to be thought of as belonging to a special branch of the arts variously known as the "mechanic" (or "practical," or "industrial," or "useful") – as distinct from the "fine" (or "high," or "creative," or "imaginative") – arts. ("Postmodern Pessimism" 242)

The etymological origins of the word capture this intertwining of thought and act, art and mechanism. It is derived from the Ancient Greek *techné* (craftsmanship, craft, art) and *logos* (ground, word, order, knowledge, reason), but it would not start to become deployed in its modern usage until "the era when electrical and chemical power were being introduced... [W]hen these huge systems were replacing discrete artifacts, simple tools, or devices as the characteristic material form of the 'mechanic arts,' the latter term also was being replaced by a new conception: 'technology'" (245–246). Marx sees here, built into the initial deployment of the word, the capacity for mediation, for visceral insulation, identified negatively by William Carlos Williams and euphorically by the architects of the Enlightenment before him, a capacity fostered by technology's

> relative abstractness, as compared with "the mechanic arts[." It] had a kind of refining, idealizing, or purifying effect upon our increasingly elaborate contrivances for manipulating the object world, thereby protecting them from Western culture's ancient fear of contamination by physicality and work. An aura of impartial cerebration and rational detachment replaced the sensory associations that formerly had bound the mechanic arts to everyday life, artisanal skills, tools, work. (248)

So *this* is technology: artefacts that are modern, complex, and insulating and, possibly, the kinds of practices that they require. It is the unreliable things created after you were born whose current totem is the computer where once it was clockwork, the steam engine, the electric grid, the nuclear plant (technology is often symbolised with power). Is this enough?

In work from a wide variety of philosophers and cultural theorists such as Bernard Stiegler (*Technics and Time*); Martin Heidegger ("The Question Concerning Technology"); Jacques Ellul (*The Technological Society*); José Ortega y Gasset (*Meditación de la Técnica*); Paul DeVore (*Technology: An Introduction*); and Michel Foucault ("Technologies of the Self") (and we could certainly go on) there have been frequent and repeated attempts to redefine or augment our notions of what "technology" actually refers to, and studies have been undertaken in the pursuit of a definition from which to begin a pedagogy of the subject (e.g., Hansen and Froelich, "Defining Technology and Technological Education: A Crisis, or Cause for Celebration?").[2] That such projects were, and are, required suggests that the trajectory of the term identified by Marx is insufficient; gesturing towards modern, complex, and insulating artefacts doesn't seem to be sufficient – there is more to technology. For instance, when archaeologists speak of the pre-human, Oldowan stone technology deployed by *Homo habilis* some 1.75 million years ago, mostly sharp edges for cutting and scraping, the first things that spring to mind are not complexity, modernity, or visceral insulation. At its worst then, our use of the term may well be obfuscating or misleading.

At the heart of the work of each of the thinkers listed above is the insight that technology is more than just material *things*, that the term might include practices, knowledge, and mind-sets as well as implements. In contemporary research, particularly in History, this trend is best seen in the deployment of the term "technological system" to replace the artefact-centred discourse that always tends to condition the study of technology:

> There seems to be a general agreement that any definition of technology must begin with material objects, but in many cases the definition extends well beyond the material core...In recent years, the

---

[2] Briefly, Stiegler is most interested in technical (as opposed to natural) objects after Aristotle and the indivisibility of the human and the technical. Heidegger, as we saw in the first chapter, deals with modern technology as an attitude, treating nature as "standing reserve." Ellul mostly concerns himself with *technique*, the skills available to a culture clustered around its artefacts, but he also sets out his own multiple point definition of modern technology (79–147). Ortega y Gasset saw technology as fundamental to human existence, and as what separates us from nature, but though he considered this to be our ideal state, one foot in the natural world and one foot in the world we make for ourselves, he also saw modern technology as of a new order and one that could be too distancing. DeVore saw his object of study as "the creation and utilization of adaptive means, including tools, machines, materials, techniques and technical systems, and the relation of the behaviour of these elements and systems to human beings, society and the civilization process" (xi). Foucault, meanwhile, focuses on the immaterial, considering the four "specific techniques that human beings use to understand themselves," technologies of production, sign systems, power, and of the self. Hansen and Froelich present prior discourses, but largely end up leaving the final issue of definition to their readers.

meaning of "technology" has been broadened as historians have come to favour the "technological system" rather than the "machine" or the "invention" as the basic unit of analysis. (Williams, "Political and Feminist..." 218)

Technological systems draw infrastructure, economic, political, and social concerns into the discussion, moving beyond the individual artefact to the web of influence in which it sits, a web, it is argued, that is required in order to understand the nature of the artefact (i.e., to a greater or lesser extent it is the sum of its relations, an idea that will be explicitly challenged here and in Chapter 4). Even though there has been this move away from artefacts as the definitive subjects of the history of technology, Marx takes care to remind us that "[a] system is 'technological'...only if it includes a significant material or artifactual component" ("Postmodern Pessimism," footnote 245). Whilst I will go on to argue that it need not be necessary to define technology by the material artefacts around which the term is often arrayed, the requirement of some physical thing is absolutely the canonical opinion in every level of the discussion of technology.[3] The principle problem with "technological system" as a more nuanced alternative, however, is that you cannot simply point to one and they are hard to describe, requiring books and papers (at least) to outline. This seems to me to be a loss, and we might yet save the word "technology" so that an academic usage can allow us to retain the ability to talk about specific items relatively independent of the conditions in which they sit. As the sociologist Bruno Latour puts it, "even though what historians call 'technological systems' do exist on the local level, they are no more made *of* technology than law is made *of* law or religion *of* religion" (*An Enquiry* 212). This is not to say that context isn't important, but it also seems crucial to be able to say "a technology" and refer to a particular thing in the world. A "technological system" might be the right term for the extensive and influential surrounds of an artefact, but it doesn't point us towards the phenomenological experience of using *this thing* and its classification as a species of object unless we have a sensitive definition of technology to begin with. A cultural aspect may well be one criteria which defines a technology as what it is, but it shouldn't be the only thing.

---

[3] Though this position is certainly not without its detractors, for example, Foucault's immaterial technologies in "Technologies of the Self," or Pitt's work in *Thinking About Technology*.

2.2 Further failures of existing definitions

In essence, then, the internal contradictions, the problematic consensus on the physical, and a troubling distinction between more "natural" simple technologies and the complexities of modern artefacts all point to a need for further work in this area. Let's start with an initial common-sense definition of technology and then build out in order to try and neutralise its weaknesses.

In order to cover its popular use, a basic definition of technology has to encompass the entire spectrum of human tool use, from stone blades to computers and with a suggestion that the latter is somehow *more* technological, an amplification of the former. Such a definition might say: Technologies are the artefacts onto which we offload tasks in order to reduce our expense of time or effort, and humans have proved themselves uniquely suited to their invention and use. "A technology" is an artefact with which we interact in order to accomplish a task that we could not by ourselves, for example a hammer for hammering, a car for driving, or a computer for its myriad uses. I would take this to be a fair starting point as a contemporary definition of technology that isn't overly distracted by the new or complex (and still commits to physicality).[4]

When Heidegger asked "The Question Concerning Technology," he also began with what he saw as a common-sense definition:

> We ask the question concerning technology when we ask what it is. Everyone knows the two statements that answer our question. One says: Technology is a means to an end. The other says: Technology is a human activity... [This] definition of technology is indeed so uncannily correct that it even holds for modern technology, of which, in other respects, we maintain with some justification that it is, in contrast to the older handicraft technology, something completely different and therefore new. Even the power plant with its turbines and generators is a man-made means to an end established by man... [T]his much remains correct: Modern technology... is a means to an end. (312–313)

Aside from introducing the matter of complexity (which he rules out, in any case, as significant to a "standard" definition), Heidegger gets to the

---

[4] One immediate complaint might be "what distinguishes a technology from a tool under this definition?" I agree. It's a real issue and one that I hope to go on to solve here.

same point from common-sense here: technology enables; technology is a human activity. It is important to note that even when he progresses away from this original definition, Heidegger keeps technology at the heart of human experience and as providing a "means to an end," a method for getting things done (it's just that, for Heidegger, that end becomes an increasingly impoverished interaction with the world even as it is forced to give up more "useful" supplies). So here lies our starting point.

It's important to ask why we might need to abandon, as Heidegger did, the definitions of technology that we already have which use variations on the above assertions as their base. In short: it is essential because such definitions are not specific or sensitive enough to deal with the complexity, variety, and intimacy of our interactions with the most significant items that extend our abilities. Some things are blithely referred to as technological, a supercomputer or particle collider say, when most of us would encounter them in much the same way as we would a worn-down inscription on an unfamiliar monument – we are dimly aware that there is a meaning attached to the object, that there is information others may have gleaned, but to us it is inaccessible and so smoothly excluding as to eventually be ignored as an inert facet of the world. This doesn't seem to describe our simply understood and precise interactions with a hammer or a knife in the actions for which they were designed, and yet these too are certainly technologies, the technologies from which all of our current interactions have emerged. A nuanced definition of the term should be able to account for the experience of the knife and of the collider, to account, I will argue, for both initiate and expert use, and to recognise each individual's unique encounter with a thing. As the philosopher David Rothenberg has argued (though to different ends) a "successful explanation of technology should not blur saxophones and motorcycles, nuclear power plants and ball point pens, all into one wrong turn in the story of our species" (xiv).

A second, and equally important, reason for the clarification is that this chapter, and this book as a whole, argues that the expert use of artefacts is intimately a part of what it means to be human, that we use equipment in order to apprehend the world, to define our place within it, as extensions of our bodies and the cognition they are entwined with. Any suitable definition of technology should also be able to account for, and to emphasise the centrality of such activity in our lives, and render ideas such as "technology is unnatural" untenable by default.

In order to avoid using the term "technology" unspecifically, Heidegger's term "equipment" (from *Being and Time*) is useful in its simplest sense: an item for getting something done – "[w]e shall call those entities which we encounter in concern '*equipment.*' In our dealings we come across equipment for writing, sewing, working, transportation, measurement" (97).[5] This term has, at its core, the notion that we encounter such items "in concern," purposively, for "our most basic way of understanding equipment is to use it" (Dreyfus 64). This commitment to *use* must remain in any definition of technology: a technological interaction cannot occur by accident. But not all equipment are technologies.

2.3 A new definition

In establishing my own definition of technology, one sensitive to the issues raised above and to those of Chapter 1, I want to build on the broad definition outlined by Pitt in *Thinking About Technology*. Pitt sees technology simply as "*humanity at work*... [T]his definition allows us to make the distinction... between the tools and their use. The tools themselves are not the technology; it is the use to which they have been put that marks out a technology, and it is people who do the putting to some use for some purpose" (11–12, emphasis in original). I want to nuance this definition, but I also broadly agree – there is no essence of technology within an artefact, but there are ways of using things that can cause us to encounter them as something we may want to refer to as "technological" (for me the technology is not the use of the equipment, but the way that certain uses cause us to encounter the equipment as something distinctive).

As a preliminary definition in line with Pitt's, one that prefigures the discussion to come, I want to say that "technology is humanity at *repeatable and expert* work." I will argue that this distinction (repeatable and expert) is vital to understanding how broad categories of use (e.g., initiate and experienced) differ, and how the process of moving from one mode to another, in whatever area, has common features.

Also following Pitt, I want to extend my own definition beyond classes of objects and instead look to the effects of deployment. Pitt suggests that we

> include as many dimensions of technology as possible. One way to accomplish this is to expand our account beyond the more standard view of tool-as-mechanical-mechanism to tool-as-mechanism-in-general...Some of humankind's earliest and most important tools

---

[5] For more on Heidegger's use of the term "equipment" see Hubert Dreyfus *Being in the World* 62–64.

were the social structures it devised or evolved for establishing order and protection. Surely these need to be included in a good account of "technology." (10)

So too, my definition will be broad, focussing on what makes equipment useful, and what uses, what "mechanisms," might count as being technological and why – in this way "technology" won't be restricted to describing physical things. I therefore use "equipment" in a way that is agnostic to materiality – I will use the term to describe things that we encounter in use, but, as Pitt notes, not everything that we use is physical.

I will tend to use the archaeological/historiographical/material cultural term "artefact" to describe a physical object produced or affected by human labour, but we can also appropriate Julian Huxley's "mentifact"[6] to describe equipment that exists as a virtual output, a useful object created in, by, and for the mind. Mentefacts won't be our primary concern here; due to this work's interest in e-reading and embodiment, and also the dominant discourse of technology as physical things, I will predominantly focus on artefacts, but the definition of technology developed below is intended as equally applicable to both classes of equipment. Particularly to begin with, to make the topic as approachable as possible at first pass, I will concentrate on contemporary human interactions with objects most people would likely readily define as being technologies under the common-sense definition outlined above, artefacts such as tools, weapons, computers, cars, and e-readers (with the latter continuing to be our model case). Later on, however, we will move on to interactions that may prove more controversial to define as such.

I will defend my reasoning for each of the following four criteria, but first I would like to simply outline what I take to be the minimum effects an interaction with equipment has to cause or enable in order for it to be considered a technology:

- Technologies EXTEND our means or ability to accomplish tasks.
- Technologies are COMMUNAL, existing only in communities of users.

---

[6] In his 1955 editorial for the *Yearbook of Anthropology*, Huxley described a trifecta of artefacts, mentifacts (I'll revert to the UK e), and sociofacts. Huxley thought of mentifacts as a culture's ideas, beliefs, and structures of knowledge; "sociofact" describes specifically interpersonal interactions and social structures (and won't be our concern here).

- Technologies are able to become, if only temporarily, skilfully INCORPORATED into our embodied cognition.
- Technologies have an effect on their users, they are DOMESTICATING.

So: Extension, Communality, Incorporation, and Domestication; it is on these terms that I will discuss what our most important artefacts do to and for us.

2.3.1 Extension

For an artefact to be considered a technology, that is, equipment at the heart of a technological interaction, it must Extend our capabilities. This assertion is deeply rooted in any definition of technology, and also in Heidegger's notion of "equipment." When we approach an object with a concern or purpose it is most often because we are able to achieve something through our dealings with it that we could not otherwise accomplish by ourselves, or the interaction saves us time or effort (again, this is about achievement, of a faster speed or a less tiring process).

The media theorist Marshall McLuhan suggested that technologies are augmentations (Extensions) of our basic apparatuses: the phone, for instance, augments the mouth and ear; the television the eye; the clothing the skin; etc. (*The Medium* 31–40). All technologies Extend some aspect of ourselves in this way, whether relatively trivially such as a shoe Extending the range of abilities achievable by the foot (covering rough terrain, sports use, etc.), or profound, such as the spear's Extension of the hunter's arm which allowed for an immense shift in culture and every change that was entailed by such a shift. Following on from an idea established in Chapter 1, Andy Clark also notes that this capacity for Extension, or at least our panoply of Extensive interactions, is a uniquely human trait: "We alone on the planet seem capable of creating and exploiting such a wide variety of action amplifiers, ranging from hammers and screwdrivers, to archery bows and bagpipes, to planes, trains, and automobiles" (*Supersizing* 157).

Extension through equipment is the most fundamental aspect of defining technology, and it is where many definitions stop – for most people who consider it, Extension *is* technology. I want to argue, however, that truly technological interactions require more.

> *Extension: The e-reader and the codex Extend our ability to store and retrieve information and to communicate from one to many across gaps of distance and time.*

2.3.2 Communality

For an artefact to be considered a technology, that is, equipment at the heart of a technological interaction, it must exist in a Community of users, prompters, and refiners. This criteria is indebted to the work of "technique" philosophers, such as Ellul, who consider technology to be defined by the modes of thought that it prompts in a society, something also captured (at least in part) by the historiographical term "technological system." I would like to suggest, however, that the best way to understand technology lies not in its either being the artefact(s) under consideration, or its/their cultural entailments, but instead a blend of the material, the personal, and the social as they affect a single user's particular interaction with a single (perceptual) unit. A technology does not exist in isolation; in order for its effects beyond Extension to come to pass, an artefact requires prompts from outside of the individual, and in a wide variety of ways our experience of any object is affected by variations of received wisdom (something that we will consider in more detail in Chapter 3).

The criteria of Communality is important because it acknowledges the ways in which our perception of artefacts, and the potential range of our intentionality when deploying them, are always at least partially socially constructed – we don't encounter artefacts neutrally (and, as we will go on to explore, artefacts can never be wholly neutral themselves, even if they aren't the dramatic shapers of our thoughts and actions envisioned by hard line technological determinists). "A technology" is always a shorthand for a set of specific community-structured interactions clustered around an artefact which is itself an encounterable yet mutable conglomeration of discrete parts.[7]

Under the criteria of "Communality" comes the notion that technological interactions are always motivated and structured by communal pressures. I have intentionally chosen the term "Communality" over "Culturality" as "culture" is a loaded term, at the least implying fairly strict delineations and histories; "community," on the other hand, can be any grouping of any pedigree, from societies down to subcultures, small groups, or even pairings. Any of these different sized communities can exert cultural effects, where "cultural" refers to the kinds of pressures

---

[7] We'll pick up on this idea in the next chapter with the idea that every artefact is experienced as a "gestalt," a plastic amalgamation of component elements that we encounter in varying combinations dependent on our skill and culturally-conformed intention. It is important to state from the outset that the focus on skilful use and individual exploration prevents this from being a wholly social constructionist/constructivist model, though it is clearly indebted to the constructionist/constructivist insight that social norms and socially defined truths do have an impact on experience.

a community exerts on its members. By avoiding the term "culture," if not "cultural," we can account for the existence of newly developed technologies in niche groups such as computer hackers, circus performers, Formula One teams, or quantum-physicists. Later in this chapter I'll argue that the equipment produced by such groups would not be encountered as technological by the wider societies in which such groups exist, at least in part because of the absence of communal pressures and support.

To further define Communality, let's consider its role in the inception of basic tool use. Merlin Donald notes that "[i]nnovative tool use could have occurred countless thousands of times without resulting in an established toolmaking industry, unless the individual who 'invented' the tool could remember and re-enact or reproduce the operations involved and then communicate them to others" (*Origins* 179). When a single being, through trial and error, was able to crack open a hard-shelled nut with a rock for the first time, they did not create a technology, and their interaction with the rock was not technological. Technologies emerge when work is driven by cultural pressures and enters into a community of beneficiaries or stakeholders. It is in a community that the drive to repeat such tasks is refined (the positive reinforcement of being able to access food others can't in this instance, thereby achieving hierarchical or survivable advantage) or through a community's inspiration to attempt experiments in the first place and in the future.

Beyond motivation and support for new creations, it is an enduring community that provides the next generation of tool users with the practices and skill sets that it would be impossible to accrue even through a lifetime of trial and error:

> humans link with a vast and diverse cultural matrix in early infancy and profit from the rich storehouses of knowledge and skill that have accumulated in our cultural memory over many millennia...The human brain is the only brain in the biosphere whose potential cannot be realized on its own.[8] It needs to become part of a network before its design features can be expressed. (Donald, *So Rare* 150 & 324)

---

[8] Extending Davis' dismodernism, I suspect that most brains require environmental support of some kind (cf. the later discussions of enactive and situated cognition in Chapter 4) and that many communal animals require others in order to be fully realised. Jacques Lacan, for instance, in his discussion of the "Mirror Stage" in childhood development, mentions the case of the development of the female pigeon: "it is a necessary condition for the maturation of the gonad of the female pigeon that it should see another member of its species, of either sex; so sufficient in itself is this condition that the desired effect may be obtained merely by placing the individual within reach of the field of reflection of a mirror" ("The Mirror Stage as Formative of the Function of the I..." 504). But the point still holds that such enculturation is distinctively potent (and particular) for human cognition.

This assertion is crucial to the thesis of Donald's work in *A Mind So Rare* – his placing of consciousness as the driving force of human evolution is supplemented by the idea that it is as communal agents, able to offload skills and problems into a "cultural memory" to be drawn on as needed, that we were able to emerge as skilled equipment users, technically adept and able to adapt rapidly to whatever environmental pressures we might face. Donald states that we are unable to reach capabilities that most of us would consider to be fundamental to our natures without some process of enculturation; there are elements of our minds that will not come "online" until we are embedded amongst cultural pressures.

> Our cultures invade us and set our agendas... [o]nce we have internalized the symbolic conventions of a culture, we can never again be truly alone in semantic space, even if we were to withdraw to a hermitage or spend the rest of our lives in solitary confinement. Big Brother culture owns us because it gets to us early. As a result, we internalize its norms and habits at a very basic level. We have no choice in this. (*So Rare* 298–299)

This then begs the question: can someone create a technology if they retreat away from Communal interaction to live alone? If someone working in the woods for long stretches, for months or years, comes up with new equipment for accomplishing a task, a new tool say, can their interaction with it be thought of as technological; will it become a technology for them where no community exists to aid in its standardisation? I believe so, because that person, in this special case, brings their community with them, much as Donald described in the above quotation. Shaun Gallagher and Dan Zahavi similarly see this community effect shaping our perception of objects: "something affords me possibilities, only because I have seen some of those possibilities actualized by others... [A]lthough they may not be perceptually present, they are potentially and implicitly involved in the very structure of my perception" (*Phenomenological Mind* 103). The influence of other users from our communities becomes built into who we are and how we encounter things in the world, and this is something that we will always carry with and within us.

I don't believe, however, that a technological interaction is possible from a person with no prior knowledge of a community. The experiences of so called "feral" children,[9] for instance, have told us much about the brain's dependency on cultural forces for it to reach its full symbolic capacity: "Socially isolated humans do not develop language or any form of symbolic thought and have no true symbols of any kind. In fact, the isolated human brain does not act like a symbolizing organ ... It is apparently unable to generate symbolic representations on its own. It does so only through intensive enculturation" (Donald, *So Rare* 150). The same can be said of a child who is simply too young to take on board the pressures of its own limited milieu; cut off from Communal pressures technological interactions are impossible. But such interactions can become standardised in the gaze of an *internalised* community that continues to shape the mind of our creator in the woods. This could only be empirically evidenced by the types of equipment that would be produced by people of different cultures entering into isolation: such objects would, presumably, reflect the cultures from which those persons have been exiled, showing what values have been instilled, what is most important, the preference of efficiency or craft, etc., but without the dramatic ethical breach required by systematic study this must remain speculative. For present and preliminary purposes it will instead suffice to say that community forces, both immediate and internalised, shape (but do not wholly define) the ways in which artefacts are encountered and privileged, what interactions might occur with or alongside them, and the drive to produce such interactions in the first place. Communality as a criterion, then, is the requirement only that a technology cannot exist as such in the mind of just one person (we will shortly consider another branch of interactions with equipment, "devices," which aren't technological, and we will see that one of their defining characteristics is their occurrence comparatively independent of a community's cultural pressures). The standardised use of an artefact in a community, then, is the second "mark" of the technological.

---

[9] For more on feral children see Michael Newton's *Savage Boys and Wild Girls: A History of Feral Children*; Adriana S. Benzaquen's *Encounters with Wild Children: Temptation and Disappointment in the Study of Human Nature*; and Serge Bordered's *L'Enigme des Enfants-loups* [The Enigma of Wolf-children]. Russ Rymer's *Genie: A Scientific Tragedy* further tells the story of a girl trapped and isolated by her father for the first seven years of her life and her ensuing exploitation by developmental scientists keen to test their theories. There is, sadly, a thread of tragedy, exploitation, and lack of compassion through a great many cases of un-(der)socialised children and adults.

Communality: E-readers and codices exist in communities of users of every scale, from international groups of millions or billions down to small groups of specialists. These communities structure users' interactions with these artefacts, determining to some extent the experience of the encounter and the uses to which the equipment is put.

2.3.3 Incorporation

For an artefact to be considered a technology, that is, equipment at the heart of a technological interaction, it must have the capacity to be Incorporated into the user's sense of themselves, their range of perceived abilities, intentionalities, and capacities, through skilful use. As we will go on to explore, this act of Incorporation takes many forms such as muscle memory, heuristics, skillsets, techniques, and other standardised approaches to tasks. Due to this book's particular interest in embodied technological relationships, this chapter will focus on the potential for Incorporating physical artefacts into the user's body schema, but the criterion of Incorporation is predominantly about a reliable alteration of the user's conception of her potential for high quality action.

When we encounter an artefact as equipment with which to accomplish a task, then we can refine that usage, standardise it, and go some way to making its skilful deployment automatic. When we first use a saw, for instance, the jarring back and forth as the teeth catch in the grain are a world away from the expert carpenter's easy push and draw – a path of learning based on at least somewhat attentive refinement must occur in order to pass from one towards the other, with a large expense of effort and time required to reach truly expert practice.[10] I raised the issue earlier that an artefact might be encountered as a technology by one user, and not by another, and the criteria of Incorporation marks the most obvious example of that break. Sometimes an artefact isn't encountered as a technology because the prospective user doesn't recognise that it might Extend her abilities. Sometimes an artefact isn't encountered as a technology because it is somehow encountered in isolation, outside of a Community of stakeholders. But, most commonly, an artefact isn't encountered as a technology because the process of refining use hasn't occurred – the user feels the equipment as separate from herself, and this can apply just as readily to e-reading, car driving, and machine-gun firing as it does to the interaction of carpenter and saw.

Certainly both the master and the novice user are extending their range of considered abilities, neither could hope to cleanly break the

---

[10] *10,000 Hours* according to Malcolm Gladwell's popular estimation.

wood without the equipment, but the modes of interaction are so significantly different from one to the other that we should consider one to be technological and the other not – we can say that the novice's experience is not technological, and that she does not encounter the saw as a technology. This seems to be recognised in our hesitancy, historically, to describe apprentices by the title of their craft – experts in crafting wood are carpenters, it is a part of their being; apprentices are apprentices, they are yet to bond the activity to their souls. What distinguishes the expert user is that the saw feels like a true extension, a rich addition, not a numb prosthetic. It is not just that the tool opens new possibilities, but that those new possibilities are encountered by a body that is *augmented* by the equipment. To make it clear: the novice or inexpert user encounters, in some significant way, *a different artefact* to the expert who brings the object "on-board" with their body (the ontology underpinning this claim will be taken up more substantially in Chapter 4).

The equipment is properly thought of as a part of the body during expert use, a "soft-assemblage" – an entanglement where components temporarily come together, and are just as easily separated, but when so joined allow for a function far greater than either component of this new machine (for sawing, or driving, or computing) in isolation. Andy Clark describes such a union between user and technology as "transparent":[11]

> Transparent technologies are those tools that become so well fitted to, and integrated with, our own lives and projects that they are ... pretty much invisible-in-use. These tools or resources are usually no more the object of our conscious thought and reason than is the pen with which we write, the hand that holds it while writing, or the various neural subsystems that form the grip and guide the fingers. All three items, the pen, the hand, and the unconsciously operating neural mechanisms, are pretty much on a par. And it is this parity that ultimately blurs the line between the intelligent system and its best tools for thought and action ... There is no merger so intimate as that which is barely noticed. (*Natural-Born* 28–29)

Incorporation therefore further troubles the idea of technology as an insulating layer between us and the world. If a tool is brought on board,

---

[11] Any reader familiar with Heidegger's work will recognise Clark's allegiance here with the mainstream interpretation of Heidegger's tool analysis from *Being and Time*, in particular his discussion of readiness-to-hand; we'll discuss this in detail shortly.

made transparent in Clark's terms, then our encounter with the world *through it* will seem akin to that with the hard assemblage of the components of our bodies.

In order to understand how the body can achieve such a synthesis, and to defend its use as one of the defining traits[12] of a technological interaction, I will take some time to explore its corollaries in several academic discourses, principally from Philosophy and contemporary Cognitive Science. Incorporation is potentially the most controversial of the four criteria outlined here, but it is vital to what this chapter, and this book as a whole, is trying to argue: Technologies are indeed a class of thing, but those things are constrained as such only by the perception of their users – if a user experiences an artefact as a technology then that is what it is, if they do not then it is not. My task is to try and outline what common features occur that, when experienced together, result in a mode of interaction that we might productively call "technological," recognising it as one of the great (and neglected) shaping experiences of what it means to be a human being. The four criteria are intended only to define the terms of an experience which might be appreciated as technological, and the criteria of Incorporation eliminates a substantial amount of artefacts from that class "technology."

2.3.3.1 Body schema

To begin to explore Incorporation I'd like to first introduce the notion of a "body schema," a classical neurological paradigm which has been reinvigorated by contemporary research. The body schema is essentially the mind's typically unconscious model of the material body's external boundaries and postures:

> it has been found that besides proprioception[, the awareness of one's limbs in space, particularly focussed on feedback from joints], other sensory modalities (typically somatosensory[, sensory reception from skin, muscle, bone, internal organs, and cardiovascular system,] and visual) are crucial to its construction... [S]ingle-neuron recordings in the monkey brain have changed the vision of a 'purely perceptual' construction of a body map in the brain towards a more multicomponential, action-oriented one. In this view, multiple fronto-parietal networks integrate information from discrete regions of the body surface and external space in a way which is functionally relevant to specific actions performed by different body parts. (Maravita & Iriki 79)

---

[12] As we'll go on to explore, "trajectory," rather than "trait," might be the more appropriate term.

The neuropsychologists Angelo Maravita and Atsushi Iriki here outline how that model of the gross bodily form is created and constantly updated in a dynamical relationship with the surrounding environment to be acted in. Rather than a representation formed from idle sensory perception, the body learns about itself by acting, becoming aware of how its surfaces and forms functionally relate to an immediate environment.[13]

The body schema is made up of our cumulative, emergent, and on-going interactions between our physical form and the environment, and it can be used as both a preconscious guide to our actions and as a conditioning force for our conscious perception of ourselves.[14] When we use a hammer there is a large amount of processing that occurs in order to allow us to accurately position the face of the tool for a good strike, and during truly technological interactions with physical equipment the hammer is brought in as an aspect of the body schema to be processed rather than an addition to it. This can occur both consciously and preconsciously, for both body schema and phenomenological body image – "it's like it's a part of me."

When the expert carpenter uses her saw it becomes softly assembled into her body schema; when we say that it Extends the abilities of her

---

[13] Later on we'll encounter Anthony Chemero's anti-representationalist *Radical Embodied Cognitive Science*. I have a large amount of sympathy for Chemero's project, as will be seen, but the body schema comes close to being a classical representation (i.e., something stored in the mind that the brain works on rather than interacting with the world directly). The body schema as outlined by Maravita and Iriki, though it clearly has representational elements, seems different enough to remain radical in Chemero's sense – it rests on an on-going relationship between the body in action with the world. I am not necessarily committed to anti-representationalism here, and the body schema's ontological status need not particularly concern us, but if it is a representation then I would take it as a blow to the anti-representationalist stance due to its usefulness and apparent testability. I suspect, however, that its dynamical and preconscious nature makes it something slightly different.

[14] For a survey of the confusion in distinguishing between "body schema" and "body image" see Shaun Gallagher (*How the Body* 1–24). Gallagher identifies the body schema as

neither a perception, nor a conceptual understanding, nor an emotional apprehension of the body. As distinct from body image it involves a *prenoetic* performance of the body. A prenoetic performance is one that helps to structure consciousness, but does not explicitly show itself in the contents of consciousness. (29)

A body image, in contrast, is a conceptual understanding of the body in space and separates us from our environment, whereas a body schema acts in concert with, or even incorporates aspects of that environment. (38)

arm we can also say that the dynamic model of the assemblage of the arm alters to include the tool, and this new assemblage necessarily has new qualities, and therefore new abilities, over the previous skin, bone, and flesh "hard" assemblage of her pre-extension body. Ong hinted at something similar when he noted that "intelligence is relentlessly reflexive, so that even the external tools that it uses to implement its working become 'internalized,' that is, part of its own reflexive process" (81) – the particular character of human thought relies on our ability to treat useful things like they are a part of us (indeed to make it so).

When the carpenter uses her saw she no longer has to think about it, any more than we must think about the pressures and tensions in our hands and arms when we perform the terrifyingly complex action of reaching out and picking up an object.[15] The same cannot be said of the interactions of the novice user who is cripplingly aware of every aspect of the interaction and the challenge of linking each discrete event together. Part of the teaching during an apprenticeship will be the repetition of motions, over and over and over again, until conscious contemplation need no longer occur. This is a process akin to one of our first acts of complex motor control, that of learning to walk in infancy: the body itself must learn how to co-ordinate its assemblage of elements in such a way that it functions, and unconsciously, as a seamless unit; it must form its body schema during active use. As Clark describes the bolting on of additional artefacts,

---

Due to the richness and invisibility of this self-conception of the body in fluid action, I'll continue to deploy and investigate the concept of body schema throughout, though we will turn, particularly in the next chapter, to some instances of learning which do require the explicit conscious awareness of a body image delimited by skin.

[15] Clark outlines this mechanism:

Posterior parietal subsystems...operate unconsciously when we reach out to grasp an object, adjusting hand orientation and finger placement appropriately. The conscious agent seldom bothers herself with these details; she simply decides to reach for the object, and does so, fluently and efficiently. The conscious parts of her brain learned long ago that they could simply count on the posterior parietal structures to kick in and fine-tune the reaching as needed.

Related to the fluid use of the saw, he continues:

In just the same way, the conscious and unconscious parts of the brain learn to factor in the operation of various non-biological tools and resources, creating an extended problem-solving matrix whose degree of fluid integration can sometimes rival that found within the brain itself. (*Natural-Born* 31–32)

The trajectory towards such fluid use is a process that I will go on to call "technologising."

what is special about human brains, and what best explains the distinctive features of human intelligence, is precisely their ability to enter into deep and complex relationships with non-biological constructs, props, and aids...We have been designed, by Mother Nature, to exploit deep neural plasticity in order to become one with our best and most reliable tools. (*Natural-Born* 5–7)

"Deep neural plasticity" describes the body schema's constant updating and refining of itself, the deployment of novel strategies when inbuilt routines fail us, the brain's ability to "rewire" its pathways in order to perform new tasks rather than relying on the ossified channels of past successful strategies, and the acceptance of external equipment into its mutable being. It is this drive to become one with our artefacts that we see in apprenticeship, the forcing of the painfully conscious into unconsciousness. This, then, forms the logical basis of the second important assertion for the criterion of Incorporation: technological use is skilful use. Skilful use is at least partly unconscious; in order to differentiate between the experience of the novice and the expert we will say that experts encounter the equipment that their expertise is directed towards skilfully; thus experts experience a technological interaction with their equipment, an unconscious interaction, whereas the novice does not – a saw is not a technology to the initiate because it is so frustratingly present.

Incorporation, therefore, is the skilful deployment of equipment for Extension that has been rehearsed into invisibility. "Whether we are learning to weave, manufacture tools, or cook food, we must learn a set of basic action sequences, generalize them, and rehearse them until they become second nature" (Donald, *So Rare* 264). This, then, goes some way to explaining why the Large Hadron Collider is not of the same order of objects as my mobile phone: The criterion of Incorporation, perhaps unintuitively, acknowledges that as I do not experience the Large Hadron Collider skilfully, that is invisibly, then I cannot encounter it as a technology; I am forced to encounter it as another class of thing.

2.3.3.2 Becoming at one

The idea of invisible or transparent tool use has been discussed in various ways across various disciplines (though, to my knowledge, no one has previously suggested that it should be considered as a defining element of what constitutes a technology). I would like to consider some of these prior instances in order to better justify the criterion of Incorporation before looking at some further implications of this requirement.

Let's first take a literary example, though one which also draws on the evidence of Neuroscience. David Foster Wallace's treatment of human reaction times in tennis, detailed in his essay for *The New York Times* on Roger "Federer as Religious Experience," is such a perfect description of the process of Incorporation that it is worth quoting at length:

> Mario Ancic's first serve...often comes in around 130 m.p.h. Since it's 78 feet from Ancic's baseline to yours, that means it takes 0.41 seconds for his serve to reach you...This is less than the time it takes to blink quickly, twice...The upshot is that pro tennis involves intervals of time too brief for deliberate action. Temporally, we're more in the operative range of reflexes, purely physical reactions that bypass conscious thought. And yet an effective return of serve depends on a large set of decisions and physical adjustments that are a whole lot more involved and intentional than blinking, jumping when startled, etc.... Successfully returning a hard-served tennis ball requires what's sometimes called "the kinesthetic sense"...English has a whole cloud of terms for various parts of this ability: feel, touch, form, proprioception, coordination, hand-eye coordination, kinesthesia, grace, control, reflexes, and so on. For promising junior players, refining the kinesthetic sense is the main goal of the extreme daily practice regimens we often hear about...The training here is both muscular and neurological. Hitting thousands of strokes, day after day, develops the ability to do by "feel" what cannot be done by regular conscious thought. Repetitive practice like this often looks tedious or even cruel to an outsider, but the outsider can't feel what's going on inside the player – tiny adjustments, over and over, and a sense of each change's effects that gets more and more acute even as it recedes from normal consciousness.

Here Wallace outlines the necessity of being able to treat the racquet as an Incorporated component of the body: to consciously contemplate it would be to remove the potential for action.

The complexity of using a tennis racquet accurately and at speed requires hours of practice in order for it to be accepted into the player's body schema and refined into expertise due to the fine grain of the control required. It can be likened to using a tool to extend our reach – if we stand on one side of a room and reach across it with a long broom handle we might, with very little practice, be able to hit something like a clock hanging on the opposite wall. Flicking a light switch, however, would be far harder, taking many attempts to determine accurate heft and balance. A single returning stroke in expert tennis play is maybe comparable to using the broom handle to extinguish a candle on the other side of the room. In 0.41 seconds. The only way to achieve such results is to make the racquet as dextrous a part of the body as the hand which holds it.[16] As Wallace goes on to point out, this is the same process by which people learn to drive cars. The novice driver does not initially experience the vehicle as a fluidly manipulable technology, but slowly, as her expertise increases, she will sublimate all of the small gestures and sweeping movements that must occur in order for her to control a car and assess the driving environment, and it is only at this point that her interaction might be considered as technological under the definition outlined here, that is, when driving becomes reduced to *intention* rather than the specifics of conscious manipulation of various levers, wheels, buttons, and pedals.

Wilson, in *The Hand*, similarly writes about the act of becoming one with artefacts, and it is from the following quotation that I have appropriated the term "Incorporation":

> this phenomenon itself may take its origin from countless monkeys who spent countless eons becoming one with tree branches. The mystical feel comes from the combination of a good mechanical marriage and something in the nervous system that can make an object external to the body feel as if it had sprouted from the hand, foot, or (rarely) some other place on the body where your skin makes contact with it...The contexts in which this bonding occurs are so varied that there is no single word that adequately conveys either the process or the many variants of its final form. One term that might qualify is "incorporation" – bringing something into, or making it part of, the body. It is a commonplace experience, familiar to anyone who has ever played a musical instrument, eaten with a fork or chopsticks, ridden a bicycle, or driven a car. (63)

---

[16] In their review of the contemporary field in *Tools for the Body Schema*, Maravita and Iriki look at tool-use studies which support this notion of complexity and dexterity requiring proportional training:
Intriguingly, whilst in some studies on humans the reported behavioural effects of tool-use occurred without any specific training..., in other studies substantial tool-use training was required to elicit these effects... It might be that simple acts, like pointing... or reaching with a stick will show behavioural effects without training, whereas more complex tasks involving dextrous use of a tool, such as retrieving objects with a rake..., require some training before any behavioural effects will emerge. (84)

Again, Wilson sees the body schema as restructured to incorporate external material objects into its perceptual model; the boundary line of the skin no longer functionally applies. Whether this opens up the idea of monkeys experiencing tree branches as technologies depends on how we negotiate the second criteria of "Communality." Is there a cultural pressure for monkeys to deploy tree branches in locomotion? Would a monkey raised in isolation spontaneously deploy the species' typical brachiating swing? If not, if there is a strong cultural element to the monkey's movement through the branches, then perhaps we could also consider that to be at least potentially a technological interaction. My argument is not, unlike Heidegger's original common-sense definition, that technology is an exclusively human preserve, only that the extent to which we can bring a vast array of equipment "on-board" is a uniquely human trait; that it is not ability, but *malleability* which singles us out.[17]

In their review paper, Maravita and Iriki further consider the mechanism for the achievement of such acts of Incorporation, examining in particular the research into

> [w]hat happens in our brain when we use a tool to reach for a distant object…Recent neurophysiological, psychological and neuropsychological research suggests that this extended motor capability is followed by changes in specific neural networks that hold an updated map of body shape and posture…These changes are compatible with the notion of the inclusion of tools in the "Body Schema," as if our own effector (e.g. the hand) were elongated to the tip of the tool. (79)

The evidence presented in this review provides empirical support for Wilson's notion of Incorporation, and also for the distinct (and, as we shall see, broader) criterion of technology to which I have attached that term.[18] Maravita and Iriki begin by describing the neuroanatomical discovery of

---

[17] I haven't focused too intently on the animal use of technology, not because I don't think that it's important, but simply because humans need not be the sole owners of any of the criteria that I see as being crucial to understanding the effects of our most intimately entangled artefacts. If the criteria outlined are met, by a human, a marmoset, or a Martian, then I consider the interaction under discussion to be best understood as being technological, though I am more than happy to accept that the marmoset or Martian may not value their technological abilities as highly as we do.

[18] I've chosen Maravita and Iriki's review to discuss this evidence as it was hugely influential early in my thinking about these issues and remains a perfect starting point for anyone interested in and requiring a clear statement of the field.

"premotor, parietal, and putaminal neurons that respond both to somatosensory information from a given body region...and to visual information from the space...adjacent to it" (79). This is to say that there are "bimodal" neurons which fire both in response to somatosensory sensation (e.g., touch) experienced by a body surface, such as the hand, and also in response to visual stimulus in the area immediately surrounding that surface. Referring to two studies in particular, one conducted by Iriki,[19] the authors outline the training of Japanese macaques to use a rake to reach for a food pellet dispensed out of their (the macaques) reach.[20] "In these monkeys, neuronal activity was recorded from the intraparietal cortex, where somatosensory and visual information is integrated" (79), and the studies aimed to record the activity of the bimodal neurons in this area. When using the rake for a sustained period of time, training its use, the monkeys' bimodal neurons fired in response to visual stimulus surrounding not just the hand, but also in the area around the tool; the mind of the macaque had begun to treat the rake as a part of its body – it had been Incorporated into its body schema.[21]

Maravita and Iriki conclude from this evidence that such "expansions may constitute the neural substrate of use-dependent assimilation of the tool into the body schema...Hence, any expansion of the [visual receptive field] only followed active, intentional usage of the tool, not its mere grasping by the hand" (79–81). Here we have crucial evidence of the nature of encountering equipment in use, rather than as an uninteracted-with object: Incorporation into the body schema only occurs during interaction. A hammer held at one's side, or on one's belt, is not in skilful use, and is not therefore encountered in a technological interaction. It is only during the skilful deployment of an artefact as equipment that the experienced interaction can be considered as technological, that is, a technology comes into being only during use.

---

[19] Iriki et al. "Coding of Modified Body Schema..." and Ishibashi et al. "Acquisition and Development of Monkey Tool Use."

[20] Japanese macaques "rarely exhibit tool-use behaviour in their natural habitat" (79).

[21] Evidence for Incorporation during human tool use supports the findings of the more invasive procedures used to study the macaques. See for instance Carlson et al. "Rapid Assimilation of External Objects into the Body Schema." For a similar, though perhaps less compelling (as less invasive) example of Incorporation in humans see Berti and Frassinetti "When Far Becomes Near..." In this study a patient who suffered from near space "neglect" in the right hand side of their field of vision following a stroke (i.e., the patient could not perceive anything in their right hand field of vision which the brain coded as being "near" to (as opposed to "far from") them) was nevertheless able to perceive objects

Consider the act of throwing a stone at a seagull: this is not a technological interaction (and I doubt that many people would consider it to be so, even though it certainly functions as an Extension of human ability). But refine the action to a reliable hunting technique and suddenly the stone is an artefact at the heart of a skilful experience. It's worth noting that a thrown spear is not somehow more of a technology than a thrown stone simply because it has been manufactured. The act of creation is a needless requirement in defining technology, it doesn't tell us anything – the act of building a thing doesn't imbue or impregnate it with some technological essence to be identified. Whilst it remains true that most technologies are built by humans, a naturally occurring hammer is still a hammer with all of the same potential for skilful Extension and Incorporation. The discovery of a magic tree which grew cars instead of fruit would tell us nothing about how humans interact with cars,[22] or how we encounter cars as a class of objects – it is *use* that defines technology.

We can also think of computers in the same way as the thrown stone: stumbling around the screen; hunting and pecking at the keyboard; having icons pointed out to you; learning how to double click with sufficient accuracy and speed – these are not technological interactions.[23] But to the experienced computer user, the computer can become the heart

---

coded as "far from" them on both sides. When using a tool visible in their right field of vision the patient could extend the effects of their neglect to objects which became coded as "near" because in reach of the tool, that is, the brain Incorporated the tool to such a degree that its reach equated the arm's reach in causing the brain to code items as far or near. "[T]he tool was coded as part of the patient's hand, as in monkeys [citing Maravita and Iriki's review], causing an expansion of the representation of the body schema. This affected the spatial relation between far space and the body... [P]eripersonal space was expanded to include the far space reachable by the tool" (418).

[22] Ok, something would change. If cars were naturally occurring things then it would shape our experience of them, in the same way as a naturally occurring hammer is encountered differently to something we go out and purchase, and these are the kinds of effects, often conditioned by our community, that we'll further attend to in the next chapter. My point here is simply that it shouldn't impact on whether these items are technologies at all, or how we define them as such, though it may change other aspects of their character, including what are considered suitable uses, time spent before discarding them, etc.

[23] For the reader who has forgotten the challenge of such early use, I recommend attempting basic tech support for any new user who has managed to remain basically insulated from our culture's obsession with screens – the initial interactions with a GUI interface, mouse, and keyboard, or a dual analogue stick game controller, is a time of easily forgotten frustration, even anger.

of a technological experience. The physical artefact is not a technology on its own terms, it cannot be, any more than can the stone; it is only a technology when encountered during a specific skilful interaction. The stone is as equally capable as a computer of being encountered as a technology when the interaction is of a particular type or character.

2.3.3.3 Readiness-to-hand

Let's move away, briefly, from the neuroscientific evidence for Incorporation and instead turn to phenomenology. We have looked already at Heidegger's "Question Concerning Technology," and also his definition of "equipment," but only in a limited capacity. The use of that equipment is what is central to Heidegger's search for "The Worldhood of the World" in section three of *Being and Time*, and the skilful actuation of that use is something very similar to my own use of the term "Incorporation." Heidegger's work can tend towards the gnomic, but this section of *Being and Time*, most often referred to as the "tool analysis," comes up again and again in all sorts of discussions in Philosophy, Science and Technology Studies, and Cognitive and Neuroscience. I think that it endures, and has become something of a phenomenological meme, because it speaks to something that we know but don't always have the words for until we encounter Heidegger or one of the thinkers that he's influenced: our tools change us all the more the better that we can use them, and the better that we can use them the less that we think of them, the less attention that we pay to them; they and their influence sneak in unnoticed at the times we use them best and most cleanly.

Heidegger states that "[w]e shall seek the worldhood of the environment (environmentality) by going through an ontological Interpretation of those entities within-the-*environment* which we encounter as closest to us" (*Being* 94). Heidegger's task is to get to the nature of (the) world itself, and to our Being within it, *Dasein*, an ontological concern that he believes we can only start to approach through the phenomenological study of the items in our environments that we encounter in intimate concern. In the intimacy of our dealings with equipment, Heidegger argues, the nature of the world can begin to be revealed.

There are two ways in which we can experience objects in the world for Heidegger: (i) in a theoretical stance where we encounter Things which are "present-at-hand," available for observation, but unavailable to experience as-they-are, and (ii) in use where we can more closely encounter equipment as what it is, as "ready-to-hand":

> Only because equipment has *this* "Being-in-itself" and does not merely occur, is it manipulable in the broadest sense and at our disposal. No

matter how sharply we just *look*...at the "outward appearance"...of Things in whatever form this takes, we cannot discover anything ready-to-hand...[T]he less we just stare at the hammer-Thing, and the more we seize hold of it and use it, the more primordial does our relationship to it become, and the more unveildly is it encountered as that which it is. (*Being* 98)

It is only because things have a reality, a truth with specific features, that we can deal with them in particular ways. When we are not using the hammer it is a Thing in the world, present-at-hand, and much as we may look at it we cannot access its nature, its true constitution. When we deploy it in a task, however, when we focus not on it, but on the work to be done, we start to gain some sense of it as-it-is. It is, unintuitively, in this *unfocusing*, this "receding from consciousness" that Wallace described above, that we begin to encounter something of the hammer's fundamental nature, the hammer-as-hammer, rather than some theorised entity. It is this understanding of *invisible* tool use which is crucial to my criterion of Incorporation.

Heidegger's concern is with getting to things as-they-are; our concern, for now, with readiness-to-hand is in seeing how such "primordial" relationships describe the productive invisibility of artefacts brought into the body schema. These differing concerns, however, are both fundamentally connected in the criterion of Incorporation: in this reading, Heidegger's conception of a use-driven relationship with equipment is the very same *Homo sapiens* trait of invisible skilful use that we have been discussing.[24] When something is ready-to-hand we cease to concern ourselves with its nature; conscious consideration bars us from true Incorporation. For Heidegger this allows us to come closer to the object as, at least to some extent, what it is; for our purposes it describes the bringing of that item into our perceptual map of ourselves so that the only focus is on the work to be done, not the assemblage which accomplishes it. This understanding renders readiness-to-hand and Incorporation interchangeable, the only difference being the focus on what it primarily reveals: an aspect of the world on the one hand, or an aspect of human interaction with equipment on the other. Technologies, in short, are ready-to-hand.

---

[24] We'll later encounter a very different interpretation of Heidegger's tool-analysis from Graham Harman and his work on object-oriented ontology. The interpretation outlined here is essentially the mainstream interpretation indebted to, for example, Dreyfus' *Being-in-the-World*.

The most significant effect of adopting this third criterion is just how many objects described as technologies under the common-sense definition are suddenly ruled out as such. The criterion of Incorporation introduces individual phenomenological concerns in order to differentiate between amateur and expert use, and in so doing renders a large amount of complex equipment viewed as archetypal technologies as not being technological at all. Under this criterion, if you've never seen or heard of a mobile phone, then it is not a technology to you, you could not have a technological interaction with it; it would just be a thing in the world, not even equipment, and it doesn't make sense to think of it as of the same order of things as the other artefacts in your life that you can deploy most intimately and effectively. But through use, the mobile phone can start to function technologically: the cultural pressures of your community drive the adoption and perseverance of that use, codifying and standardising it, and when this begins to occur the range of abilities available to the ear and mouth are extended in McLuhanist fashion. When you've texted and texted and texted, and chatted and chatted and chatted, and when you feel it buzz in your pocket and you don't even think, your hands already know what to do, and they fumble the thing up to your face with the call already answered because you know, instinctively, that that's how to avoid that little delay between pressing the button and it actually connecting, that is equipment Incorporated, a technological interaction, and one which might, as a shorthand, lead us to describe mobile phones, as a category of objects, as technologies, because this kind of expert use is readily available to a significant number, maybe a majority, of users.

As Clark states, drawing on the same review paper discussed above: "[t]he plastic neural changes... emphasized by Maravita and Iriki...suggest a real (philosophically important and scientifically well-grounded) distinction between true incorporation into the body schema and mere use" (*Supersizing* 38). A happy accident of terminology makes the point very clearly here (Clark doesn't systematically use "incorporation" as an established criterion for any particular event): the work that Maravita and Iriki discuss is an effective demonstration which augments Heidegger's notion of readiness-to-hand. The macaques had to be trained into displaying bimodal functionality during tool use, but once it was there it became clear that a soft-assembled apparatus was created, truly Incorporated because it was an increasingly skilful practice. If Heidegger's assertion is that we only begin to experience aspects of the world as they are in "primordial" invisible relationships, then it is surely not just in use, but particularly in *skilful* use that this occurs. Though it is not explicitly addressed in *Being and Time,* ready-to-hand use of equipment must depend upon skilful deployment; amateur, clumsy, or otherwise ineffectual use must necessarily be a present-at-hand mode as the tool continually draws attention to itself through its refusal to function as expected or desired. We will continue to discuss the implications of skill (and its lack) more thoroughly in the next chapters.

One final assertion must be made about the criteria of Incorporation in order for it to stand up as a defining criterion of technological interactions: Incorporation can still present an interaction as weakly technological, even if it exists only in potential. Let's take again the example of someone learning to drive. The car, when the novice first enters it, is not a technology, nor are her dealings with it technological. There will be, however, an "in-between" moment in the novice's learning when the equipment that she is deploying temporarily becomes invisible, only for a second, when her hand reaches automatically for the gear-stick say, or the indicators are flicked off without a thought, when the mirrors are checked and the small manoeuvre executed without a conscious ticking through each micro-event, in short, where a particular skilful interaction occurs. At moments such as these the *potential* for Incorporation becomes apparent. It is clear that it can be achieved, it is not beyond the user, and this is the very start of experiencing the car as a technology. We might say that at this moment the equipment "trends towards" being a technology; this is a tipping point. Such users might rightly speak of their dealings as technological because the range of abilities they experience as *becoming* available to the body schema have been extended, and they have also entered into a specific set of community relations and pressures. We will return to this idea of technological trends shortly.

> *Incorporation: E-readers and codices can become invisible during use, allowing us to concentrate, for the most part, on the act of reading rather than the hand holding the equipment or the pages or screen themselves. This is not to say that these things do not affect our reading, only that they do not occupy our conscious attention if we are practiced enough for them to melt away. In this moment the body and either artefact form an assemblage for reading that is, in theory, empirically testable. Regardless of the brain's neuronal mapping, however, the phenomenological experience of Incorporated reading is one of the melting away of the apparatus.*[25]

---

[25] This assertion will be problematised somewhat in the section on "devices" below. To pre-empt that argument, a technological interaction is fragile and if interrupted the apparatus will return to our concern; this experience is not rare.

2.3.4 Domestication

For an artefact to be considered a technology, that is, equipment at the heart of a technological interaction, it must have the potential for a Domesticating effect upon the user. This is the final criterion in the trajectory of making the use of equipment into a technological event: Extension is the minimum requirement of any kind of tool use; Communities of stakeholders inspire initial use and/or refinement of technique; skilful use involves the Incorporation of the equipment into our embodied intentionality – Domestication, in turn, is a marker of the effects of such actions over time.

In its simplest guise, Domestication is the long-lasting or permanent extension of our conception of what we can achieve even in the absence of direct access to the required tool: early hammers enabled our hominid ancestors to know that they could gain access to hard-shelled food, and now cars ensure that we know that we can cover, if required, a distance far more swiftly than our biological walking gait could allow; these become things that we can rely upon. Body image and body schema are the virtual models, conscious and preconscious, of our boundaries and capacities, and any medium-to-long-term change in this perception I want to describe as "Domestication."[26] But, even on the limited timescale of the individual human life (we will expand beyond this shortly), there is more to this criterion.

Firstly, there are the relatively subtle effects of practice, where minds and bodies are tested, strengthened, and honed through repetitive and concerted interaction: the touch typist has a skill unavailable to someone approaching the keyboard for the first time; the taxi driver's senses are more alert to the particularities of the city road than those of the learner; the hunter who deploys spear or bolas has hand-eye coordination to rival the speed and agility of the quarry. For the individual user, Domestication is perhaps best represented by these changes in the brain, the increasingly well-appreciated phenomenon of "neural plasticity." Neural plasticity describes the brain's ability to change neural and synaptic pathways, and cortical mapping (i.e., what regions of the brain are responsible for), during ontogenetic development in response

---

[26] Why "Domestication"? Part of the reason for my adopting the term is its association with the dramatic changes in animals and plants as humans put them to use across our history. This process of guided evolution, where the selective pressures become partially channelled (though not always predictably) by human intention, bears a profound relationship to the human plastic adaptation to new tools that I want to continue to explore – our tools tame us.

to, for example, skill acquisition, the formation of memories, and physical trauma (e.g., re-routing pathways in the brain after a stroke). Plasticity is most apparent during early development, with the largest amount of neural connections existing at birth. As we learn, channels in the brain develop and ossify, cutting off certain connections and strengthening what has been successful (a process known as "synaptic pruning"). A large amount of plasticity remains throughout our lives, however, allowing for the continued acquisition of new abilities and the overwriting or restructuring of innate propensities.

The Incorporation of any tool must result in a physical change to an individual over time. Neuroscientific studies have demonstrated that regions of the brain dedicated to sensation can grow and shrink in proportion to their use. Alvaro Pascual-Leone and Fernando Torres, for instance, focussed on the growth of the area representing the fingertips in blind readers learning braille,[27] an example of a plastic Domestication effect in response to action. Similarly, Maravita and Iriki's review paper describes the plastic alteration of the macaques' brains in response to tool use, and Berti and Frassinetti's "When Far Becomes Near..." demonstrates a similar effect in humans (see footnote 21). Hayles also offers us another compelling example of the way this kind of plastic Domestication can manifest:

> A woman who worked on Morse code receiving as part of the massive effort at Bletchley Park to decrypt German Enigma transmissions during World War II reported that after her intense experiences there, she heard Morse code everywhere – in traffic noise, bird songs, and other ambient sounds – with her mind automatically forming the words to which the sounds putatively corresponded. Although no scientific data exist on the changes sound receiving made in neural functioning, we may reasonably infer that it brought about long-lasting changes in brain activation patterns, as this anecdote suggests. (*How We Think* 128)

Di Pino et al.'s recent "Augmentation-Related Brain Plasticity" is the most significant review to date of the existing literature and implications of the use of equipment, detailing a wide variety of plastic changes in response to tool use, augmentation, and prostheses. The paper concludes that augmentation of the body results in plastic changes in

---

[27] "Plasticity of the Sensorimotor Cortex Representation of the Reading Finger in Braille Readers."

the brain "at the cellular level... as evidenced by changes of gray matter thickness and even with neurogenesis [the spontaneous creation of new neurons] in the dentate gyrus."

Such neural Domestication, where our brains restructure themselves in response to the demands of cultural and environmental pressures, also have a corollary in the plasticity of our muscles, organs, nerves, flesh, and bones. The dramatic differences between the runner and the body builder, between the ballet dancer and the boxer, between the hands of the lumberjack and the hands of the surgeon, every body constantly shows this dramatic variability over ontogeny and expertise. This is what make humans viable as a species – we're not the strongest, fastest, or toughest, but we are the most adaptable, turning to anything in the world which might give us an advantage at any given time, and then doing our best to bring such items into the conceptions of ourselves in order to maximise the effectiveness of their use. We might therefore talk of plasticity at every scale: the micro world of adapting neurons, synapses, and cells; the meso scale of gross bodily alterations; and the macro trajectory of human evolution. This criterion leaves us with a sense of the human as being always open, reaching out, plastic, and demanding inspiration for change which often comes from expert technological use. As Clark puts it, humans are "constantly negotiable bodily platforms of sense, experience, and... reasoning... Such platforms are biologically primed so as to fluidly incorporate new bodily and sensory kit, creating brand new systemic wholes. [We are] systems evolved so as to constantly search for opportunities to make the most of the reliable properties and dynamic personalities of body and world" (*Supersizing* 37).

This seems to me to be another blow to the idea of technology being in any way unnatural. Though many aspects of our bodily experience do, of course, persist over time, it is largely due to the constraints of our physical shape at the meso-scale, rather than some ineffable "human-ness"; our "biological proper functioning has always involved the recruitment and exploitation of nonbiological props and scaffolds" (Clark, *Natural-Born* 86). Clark paints a very different picture of equipment use to that of the resisters of modern technology, the Romantics and Neo-Luddites of Chapter 1, and those who might, wittingly or unwittingly, side with them. Far from separating us from the world, our principle technological interactions bring us closer to our nature as beings adapted to function in niches within that world at all. In this regard, again, technology isn't "unnatural," it's what we do. Some of us may (legitimately) not like particular directions that we take as we are Domesticated through sustained technological use (and the related cultural changes that must necessarily go along with wide-spread embodied plasticity, for example, the rise in obesity strongly correlating with changes in diet and motility), but we cannot attack the mechanism itself as being outside of our natures. Change is an unavoidable effect of use in humans, and use, manipulation, is our default mode of apprehension; we are rarely satisfied with the present-at-hand, distanced interactions of sight, sound, smell, and contemplation when we wish to understand a thing. That a switch from page to screen must produce plastic changes in the brain and body is a certainty, but so does jogging, so does using a hammer, so does learning to read in the first place. It's not that we change, but the effects of the change that need to be considered; not a deviation from a neat and fixed human nature, but a messy shaping of that nature that we may not wish to privilege. The irony here is that it is only for users who are used to the new equipment, deploy it frequently, and likely don't experience it as problematic that the most extensive Domesticating effects occur.

Domestication, then, is my final criterion of technology. If it is not met then the interaction is not technological, but if the other criteria are in play then Domestication will inevitably follow. Plastic adaptation is the effect of the other three criteria, but it should not be ignored in its own right. If we are trying to decide whether we should consider something as being technological, or not, then one of the first avenues that we can pursue is: what does it change in its user? If the answer is hard to find then that makes for a fine indicator that a technological interaction is not occurring; technologies alter their users and require, as we will go on to see, users altered to their properties.

> Domestication: E-readers and codices both alter their users by acting as a substrate for writing, a sign system that materially affects how we think. But the artefacts themselves also Domesticate us, altering their user's conception of what they can achieve in terms of memory and information gathering and forcing us to develop the subtle muscular and sense skills of turning single leaves at a time or navigating a touch screen or system of nested menus.

2.4 Devices

With these criteria of technology established, I now want to turn to a different order of equipment, though closely related, in order to refine the idea. I want to define the kinds of equipment that we encounter in use, but that we *don't* have a technological interaction with. The word "device" seems apt for this class of useable objects, still connoting a means for getting things done, but without the implications of intimate relations that we have bound to "technology"; the word can be traced back to the Old French *devis* meaning "separation" and "division," but also "wish" and "desire."[28]

A pure device is (to borrow an IT term) a quick-and-dirty solution to a problem which exists for a single user at a single time, whereas a technology is a standardised solution to a common problem in a community, a solution which both persists for, and impacts upon, its users. An example of a device: if you were to drop your keys into a drain then you might turn to the available resources of the environment in order to retrieve them – you rush into your house, fetch a coat hanger, bend it into a crude hook, and, with a little difficulty, fish out the keys. This is a device; you interact with it as a tool, but not as a technology. It might Extend your means, but once the task is complete the problem, and your new device, do not persist and the artefact has had no long-term effect on the way that you interact with your environment. This is directly akin to the stone pitched at a seagull as opposed to the refined practice of using a spear for hunting.

Taylor also provides a helpful anecdotal example from a fishing trip where he had to improvise a club from a nearby rock:

> Sacrificing smoothness for weight, and balancing a moral need for swift dispatch against my affection for my own fingers, I used three or four medium-weight blows. These unaesthetically but convincingly spit the skull, knocking the eyes out. Things improved with the second and third trout, and the fourth was neatly sent to wherever trout go when they die (my stomach I suppose). Afterward, Keith [Taylor's fishing partner] tossed our expedient artifact back into the water, and as the blood billowed off downstream, history evaporated. Unlike the rod, hook, and line, the improvised fish whacker reverted to being just another rock, unmodified and non-cultural. (*Artificial* 45)

Taylor's fish-dispatching device took practice to use well; we can see a process of Incorporation emerging here, and it also Extended his

---

[28] There is also a similarity between my deployment of "device" here and that of Albert Borgman in *Technology and the Character of Contemporary Life*. For Borgman, a device is an item of equipment, tangible or intangible, which separates the user from the realities of the work to be done and requires no skill (e.g., central heating vs. a wood-burning stove). For my usage, device will similarly describe those instances of use which do not draw the user closer to the work, the object, or themselves, that is, instances that don't abide by the criteria discussed above.

abilities. The difference between its being a device or a technology lay in its Communal function: community knowledge may have motivated the search for the equipment, but Taylor's method had no return to a cultural existence; when the tool was discarded, when it ceased to be equipment in use, it disappeared.

The distinction between a device and a technology, I would like to suggest, is not a binary opposition, but instead an analogue scale. We can move from approaching an object as a device to interacting with it as a technology, indeed this must occur frequently: it is the only way that technological usage can arise as we can never experience any object as technological instantly, due for the most part to the criterion of Incorporation, a state which, as we have seen, requires rehearsal into unconsciousness, a task proportional to the unfamiliarity, complexity, and dextrous requirement of the skilful interaction. All technologies must begin as devices, novel solutions to particular problems, from which a process of refining and defining occurs; devices are "pre-technological" equipment that might trend towards being technologies over time.

Let's use our simple example of the coat-hanger-key-hook device and explore how it might become a technology, taking a similar generic trajectory to that which every technology must follow. The device already meets the first criterion: it Extends our means by allowing us to reach into deeper and narrower gaps in our environment than we might otherwise have access to. As evidenced by the struggle to retrieve the keys on the first "fish," however, the device is not yet Incorporated, but continued and repeated use could simply rectify this. And with all of these hypothetical opportunities to rehearse the device into an Incorporated state it is clear that key loss is a pressing problem in your Community. Seeing someone else struggling, you might fetch your key-hook and demonstrate its use, and following another successful retrieval you may well go on to discuss the implement, make suggested modifications, and recommend it to others. In this way the device can become a Communal enterprise, standardising its use. As for Domestication, our final criterion, who knows what sustained use by a community of users might cause? Most simply, no one in a culture of key-hook users would see a small gap with tantalising objects for retrieval and consider them to be out of reach; there would have been a fundamental shift in the perception of abilities available to the user, particularly if the equipment, driven by cultural norms, was frequently kept about one's person much like a mobile phone. As we've also seen, proficiency of any kind must also result in plastic changes in the brain. With the criteria met, the key-hook is now being approached as a technological artefact by its experienced users.

This may seem a flippant example, but if we consider how the expert users would approach the object – as everyday, as normalised, adapted to its use – then we can see that it is a far cry from the way that the initial user experienced it as she urgently fished for her keys. A process of "technologising" the device has occurred. Note that the material object itself doesn't need to change; it is an alteration of our perceptions of the equipment that we approach in use, and of our conception of what we can achieve with it.

This separation of technologies and devices allows us another way to distinguish between different ways of encountering artefacts, to name the distinction between the novice and expert user's experience, and to demonstrate that objects which we often refer to as technologies are often more appropriately thought of as belonging to a different class of equipment, such is our experience of them. We are better off reserving the judgement of "technology" for specifically encountered equipment because it allows us to more productively theorise a certain kind of interaction. "Device" might also function as a useful term in the rhetoric of those resistant to new equipment: in the case of e-reading, any promoter of the new reading equipment must justify why what is typically experienced as a fully technologised artefact, such as the codex, is being replaced by what many will encounter as a mere device. Though resistant readers might well be able to read on an e-reader, their dislike of the task, their feeling of its unnaturalness in comparison to the codex experience, may stem from the seeming unlikeliness (or pointless challenge) of the progression along a trajectory towards its becoming a technology.

2.4.1 Reversing the trend

I have already spoken of the "trend" towards technology in device use along the axes of increased Incorporation, Communality, or Domestication. The continual refining of skill appears to be a human drive; very few items are sustained in a culture where the most skilful users are merely "quite good" – there tends to be a cultural reward for those who can successfully deploy equipment to the highest level, whether that be guitar playing, photography, game hunting, aeroplane piloting, car driving, or baseball batting. But it is also impossible for technologies to remain entirely invisible, and now we can say that the trend in their direction might also be reversed – our experiences can move back from the technological to the "devicive." At these moments the technology ceases to be a part of soft-assembled, unfocused, skilful use and becomes, instead, available to conscious contemplation. It is not that the work necessarily ceases, only that it is no longer the sole, or at worst not even the primary focus of the user. Such times are marked by periods of mistake, intense concentration, and drops in the speed and quality of action.

Heidegger identifies three different types of event which manifest as the cessation of perceptual readiness-to-hand:

- "Conspicuous" equipment is that which is not as ideally suited to the task at hand as we had expected: "When its unusability is thus discovered, equipment becomes conspicuous" (*Being and Time* 103).
- "Obtrusive" equipment is marked by the absence of the equipment we truly desire to accomplish the work. The object that is available "reveals itself as something just present-at-hand and no more, which cannot be budgeted without the thing that is missing. The helpless way in which we stand before it is a deficient mode of concern" (103).
- "Obstinate" equipment places an obstacle in front of the work to be done, it "'stands in the way' of our concern" (103).

If we reach for a ready-to-hand pen and begin writing, but it is of the wrong colour, not black, but blue, then it is no longer a technology, but a *conspicuous* device. If we only have a pencil, and we need a pen, then the pencil, otherwise perfectly ready-to-hand, is an *obtrusive* device that ceases the work to be done. If the pen encounters wet paper that blocks the act of writing, then it becomes an *obstinate* device. Anything which causes equipment to return to consciousness causes a drop in the intimacy of the interaction, breaking the soft-assemblage into its constituent parts (typically at the boundary of skin and world), and rendering the item a thing to be contemplated in annoyance or simply as an object apart from what it might achieve – a device, not a technological artefact to be expertly used. In this way, these phenomenological definitions of the transition away from readiness-to-hand are useful analogues to the transition from technology back to device,[29] offering a preliminary taxonomy of how and why this might occur.

Recent experimental investigation into Heidegger's conception of the move away from readiness-to-hand can also offer empirical support for this breakdown of technological interaction. The review conducted by Maravita and Iriki demonstrates how objects can become incorporated in a way that we might recognise as ready-to-hand, but their work does not explicitly engage with Heidegger, and as such they do not consider

---

[29] For more on Heidegger's distinction between the three modes see the translators' footnote in *Being and Time* (104).

the reverse of the trend towards a skilful/technological interaction. A 2010 study by Dobromir Dotov, Lin Nie, and Anthony Chemero, "A Demonstration of the Transition from Ready-to-Hand to Unready-to-Hand," however, sought to lab test Heidegger's assertions. Dotov, Nie, and Chemero set out to test Heidegger's "description of the transition between ready-to-hand and unready-to-hand modes in interactions with tools... Despite widespread attention in cognitive science and artificial intelligence to Heidegger's work, this interest has remained largely conceptual... A search of the PsycINFO database on December 10 2009, found no articles concerning Heidegger that involved laboratory work" ("A Demonstration..."). The team's paper therefore marks the first attempt to empirically test explicitly Heideggerian ideas about tool use.

The experiments conducted by the team deployed a simple setup. Participants in the study were required to use a computer mouse to move an onscreen pointer in order to play a game. A blue dot would continually try and escape from a grey "pen"; the participant would try and use the onscreen pointer to "herd" the dot back into place. "What allows the participant to guide the target is that it always tries to escape away from the pointer in a semi-predictable fashion. To make an analogy to Heidegger's example, here the mouse plays the role of the handle and the on-screen pointer figure plays a role similar to that of the hammer striking face." However, "[a]bout thirty seconds from the beginning of the trial a perturbation in the mapping between mouse movement and pointer movement instantiates equipment malfunctioning. It lasts a few seconds and then the situation returns to normal" ("A Demonstration...").

The researchers made predictions in line with Heidegger's discussion of equipment: that the participant, as an experienced computer user, would "smoothly[, read 'skilfully,'] cope with the tool as... ready-to-hand," and that during the perturbation the mouse as tool would become the focus of attention and distract the user from a simple secondary task (in this case counting backward in threes from a given number). Their method for testing the first half of this assertion stems from the monitoring of a certain type of "noise," $1/f^\beta$, a power-law scaling in activity magnitude across the frequency range in the analysis of the hand-tool system over time. Citing van Orden et al. and Chen et al.,[30] Dotov et al.

> argue that $1/f^\beta$ noise found in an inventory of cognitive tasks is a signature of a softly assembled system exhibiting and sustained by

---

[30] "Self-Organization of Cognitive Performance" and "Origin of Timing Errors in Human Sensorimotor Coordination."

interaction-dominant dynamics, and not *component-dominant dynamics*. In component-dominant dynamics, behaviour is the product of a rigidly delineated architecture of modules, each with pre-determined functions; in interaction-dominant dynamics, on the other hand, coordinated processes alter one another's dynamics, with complex interactions extending to the body's periphery and, sometimes, beyond...By looking for $1/f^\beta$ noise recorded at the interface of body and tool, we address the hypothesis that, while smoothly operating an instrument, a human performer instantiates such an IDS [interaction-dominant system] spanning the extended body-tool system.

In short, skilful use of a tool will manifest itself in an invariant scaling of activity magnitude over a frequency range – $1/f^\beta$ noise. This has, in prior experiments on human subjects, been linked to interactions where it is not the components (i.e., the parts of the body and the equipment) that dictate the effects of the interaction, but the nature of the interaction *between* the components that dictates its effects. An interaction where the components themselves are subservient to the nature of their connection, then, is indicative of a "smoothly operating," that is, skilful, soft assembled system, not a collection of components, but a single active entity. Dotov et al.'s prediction is therefore technically laid out as the expectation of the presence of $1/f^\beta$ type noise before and after the perturbation, and an absence of such noise during disruption, indicating a move away from readiness-to-hand to unreadiness-to-hand[31] and, in our terms here, from a technological to a devicive interaction.

These predictions were born out with a marked trend away from $1/f^\beta$ noise during the disruption of the task and its presence during fluid use:

> The IDSs (delineated by the surrounding curves [in the image overleaf]) are... softly assembled by virtue of rich interactions on multiple scales (double-sided arrows) among the components (black dots and hammer)...They either span across (A) or do not (B) the tool (hammer). It is assumed that the black dots stand for bodily structures. ("A Demonstration...")[32]

We can see from the schematic below that the components of the body are always assumed to be treated as an invisible IDS, but during skilled

---

[31] Note that unreadiness-to-hand, rather than presentness-at-hand, refers to an item returning to conscious attention, but still being used. Presentness-at-hand, in this reading, is completely detached contemplating, theoretic rather than practical.

[32] Further detail on the schematic overleaf:

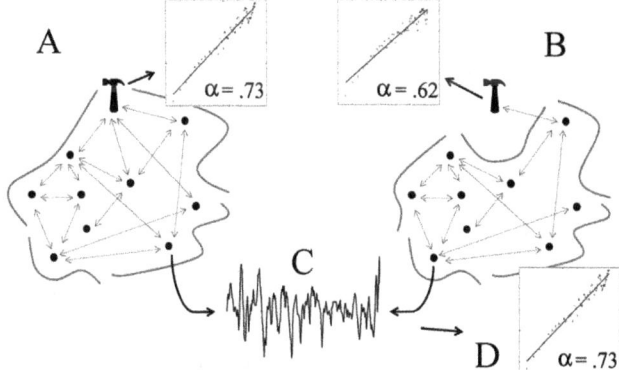

ready-to-hand use the equipment is allowed to come "onboard" and become a part of that system. This is a restructuring of the IDS in order to form a new assemblage which accommodates the tool in use. The schematic at (A) essentially shows the extension of the body schema as the virtual model of the self is updated to Incorporate the tool,[33] and this produces the smooth power-law scaling during activity that is indicative of skilful action (with or without a tool).

The conclusion of the researchers' analysis appears to support the notion of a transition from technology to device:

> We are not arguing that the flow of interaction between tool and body is reduced in magnitude [during the perturbation], just that it is reduced

---

> Customarily, one studies such systems by collecting a time series locally from the behaviour of a single point of observation (C), that is, from a single element. Next, if possible one establishes their character as an IDS [Interaction Dominant System] by searching for power-law scaling of certain statistical quantities (D)...The scaling coefficient α reveals long-range correlations characteristic of 1/f noise in the hand-tool in normal mode (A) and approaches the uncorrelated white-noise level in (B). (Dotov et al.)

Note that white-noise is associated with random interference patterns; $1/f^\beta$ type noise is sometimes called "pink-noise."

[33] As we will go on to see, Chemero is committed to an anti-representational stance that would almost certainly not allow for the seeming representationalism of something like a "body schema." The debate about representations continues (see Chemero's *Radical Embodied Cognitive Science*), and isn't my argument here, but it's important to note that if IDSs are something distinct from (or should even replace the concept of) body schema then the criteria of Incorporation also (or instead) includes tools accepted into interaction dominant systems. Incorporation describes the way in which a tool and body can come to work as one, perceptually and preconsciously, not the nature of that work.

in complexity. The mouse keeps providing sufficient local stimulation through the eyes and the sense-organs of the arm for the agent to maintain overall control over it, as when one is holding a foreign object in hand and is trying to figure out a specific property of it. (Dotov et al.)

At these moments the equipment is returned to consciousness, dropping out of an IDS, out of the body schema, in order to be observed as "conspicuous" or "obtrusive." It does not cease to be equipment, however, because we still encounter it in and for use; it is not that the work has become impossible, only that the experience with the equipment is not integrated in skilful control. This seems to ably represent a distinction between device and technology on the grounds of distinguishing between novice and expert use; the perturbation re-renders the expert user as unskilled.[34]

Failures or challenges, as Heidegger and Dotov et al. both demonstrate, are disruptive. But unexpected success, perhaps at a task we were not even attempting to achieve, is also jarring as our perceptions are again altered. We look at the tool, marvel at our connection with it (thereby breaking that connection), and it may be some time before we are able to achieve that state again. Donald, in *A Mind so Rare*, argues that the work of the conscious mind is the mid-to-long-term reflection on otherwise unconscious action, shaping that action, and shaping ourselves. Similarly, learning, as I've argued, is predominantly a conscious activity until the activity is sublimated and able to be enacted unconsciously, skilfully. A beginner juggler will rehearse the coordination of hands and clubs or balls into unconsciousness by first very consciously examining the desired movements, and then attempting to adjust their own bodies to match a mental image of approximately, and to the best of their knowledge as an amateur, what is required. At times the objects being thrown will unexpectedly match the desired trajectories and be brought to consciousness through the novelty of success, but for the most part they will be brought to consciousness as they have to be retrieved from the floor.[35] When Heidegger talks about the breakdown of readiness-to-

---

[34] In recent follow up work (Nie et al. "Readiness-to-hand, Extended Cognition, and Multifractality") the researchers re-interpreted their data with a second criterion for interaction dominant systems (multifractality), further supporting their claims.

[35] We will nuance this further in the next two chapters – the juggling balls play a role too; all artefacts affect the way in which they are encountered, and the juggler actually has to discover their expectations and plans for juggling in the undertaking of the act (theory is never enough even to begin). The example outline here is close enough to begin thinking about the transition between readiness-to- and presentness-at-hand however.

hand he is, in essence, talking about these sorts of returns to consciousness of a previously sublimated activity: "[a]lthough he concentrates on the special case of breakdown, Heidegger's basic point should be that mental content arises whenever the situation requires deliberate attention" (Dreyfus, *Being* 70). Often rehearsals in the beginning to intermediate skill levels are marked by a wildly oscillating movement between sublimation and presence, but it may be during these periods of transition that the greatest learning occurs. Little is learnt during conscious perception of the object, though a plan of action may be formed. Little is learnt during unconscious interaction; the objects, and the body manipulating them, are behaving as expected (or wished for). But during our movement between these states, as the trend from conscious to unfocused action, from device to technology, moves this way and that, we learn much about ourselves and the objects that we are attempting to tame.

How, then, might this distinction between devices and technologies be mapped onto the assertions of unnaturalness in our case study of e-reading? Through the pedagogy and practice of our early years, codex reading is refined into skilful invisibility, meets the four criteria, and should rightly be considered as a technological interaction for the typical experienced user under typical conditions. Each new codex specifically encountered, however, will slip from this pristine position as we adjust to its weight, its inflexibility, and poorly cut or set pages, etc., but within a few seconds the artefact once again seems to melt away, matching the ideal type in our minds which we identify as technological. Is it any surprise that when such expert readers encounter, or even consider, the various apparatuses available for e-reading that some of them recoil? A lifetime's work of technologising suddenly to be faced with an unfamiliar device whose manufacturer purports that it is the same, or better than what is already known – this is rightly shocking. And those early uses – radically unfamiliar weights and sizes, misplaced keys, the requirement of manuals to help navigate nested menus, no paper to smell and touch – every instance is jarring, conspicuous, obstinate, and finally obtrusive as the reader puts the thing aside and says "how distanced we've become from the world, this is unnatural." It's a lot to get used to.

2.5 Atypical technologies and the implications of the definition

With these definitions of device and technology hopefully functioning – from quick-and-dirty solutions to unique tasks to the equipment that we expertly encounter as "technological," and the infinite positions along the continuum from one to the other – I'd like to end this chapter by looking at some of the implications of these terms, firstly by extending the discussion to interactions which it may seem more radical to classify as technological. This is the vital move towards neutralising the claim that technology in general, and the digitisation of writing in particular, is unnatural.

Many writers after Darwin have become increasingly interested in what Richard Dawkins has described as a "Universal Darwinism,"[36] that is, that Darwinian evolutionary mechanisms can occur wherever selective pressures are enacted (and this will become a central theme of the final chapter of this book). What Universal Darwinism allows for is the understanding of a fundamental mechanism that can apply, with limited local modifications, to arenas of hugely different orders, scales, and subjects, for example the now famous parity Dawkins draws between the work of biological genes and cultural memes.[37] I would like to similarly suggest that the criteria that I have outlined for technological interactions with material equipment can be similarly extended onto numerous orders of interactions with synthetic, organic, virtual, digital, mental, or machinic equipment. As with Universal Darwinism, the reason for this extension is to suggest that we might better understand our interactions with these equipment if we were to observe not their metaphorical parity with an established function, but their genuinely abiding by the same rules, recognising that a difference in the substrate of their existence does not mark their function out as of another kind (a deception born of a materialist bias), but instead merely masks such similarity – the fundamental mechanisms by which they operate are agnostic to material concerns. My principle interest in the following chapters will remain with tangible things, particularly e-readers and codices, but a few brief examples below will, I hope, show the utility of the definition outlined here for broadening the debate about what technology is and does.

Let's start with a fairly well-trodden example of an atypical technology, that of literacy. James Gleick, in *The Information*, is very clear that "[l]anguage is not a technology, no matter how well developed and efficacious. It is not best seen as something separate from the mind;

---

[36] See Dawkin's influential paper of the same name and Susan Blackmore's similarly titled chapter in *The Meme Machine* for explicit references to "Universal Darwinism," but the broad applicability of Darwinian mechanisms outside of evolutionary Biology was investigated from the time of the original writings, proving particularly influential, for example, for William James' pragmatist Philosophy (see Skrupskelis "Evolution and Pragmatism: An Unpublished Letter of William James").

[37] See the final chapter of *The Selfish Gene* and Blackmore's *The Meme Machine* throughout.

it is what the mind does... [W]riting [however]... is concrete performance... [W]hen the word is instantiated in paper or stone, it takes on a separate existence as artifice. It is the product of tools, and it is a tool" (30). This, however, seems to me to be a false distinction: writing, too, can become a part of "what the mind does" – we can think in and with the act of writing in a way that is strongly related to the way that we use spoken language. Note that Gleick positions externalised physicality (i.e., an artefact outside of ourselves) as the mark of technology here, but the languages we cultivate aren't any more natural than our writing (in that we have to alter ourselves to make language "what the mind does") and the practice of writing can be brought on board just as intimately.

Jay David Bolter, in *Writing Space*, mounts an argument for considering the skill of writing itself (rather than any implement used) as being a technology and also considers the frequent resistance to such an attitude, starkly contrasting Gleick's claim:

> Writing is certainly not innate. Yet writing can be taken in and become a habit of mind. What is natural seems more intimately and obviously human. For that reason we do not wish to dwell on the fact that writing is a technology; we want the skill of writing to be natural. We like our tools and machines well enough, but we also like the idea of being able to do without them. Putting away our technology gives us a feeling of autonomy and allows us to reassert the difference between the natural and the merely artificial. (36)

Here, Bolter again reasserts the received dogma of material technology as being unnatural, with "tools and machines" being able to be "put away," allowing for a return to some natural, unaugmented state. Writing for Bolter, however, is both a technology and an element of a hard-assemblage with a biologically discrete user; it can be "taken in" as a "habit of mind." Bolter's point is significant for our discussion here: the drawing-closer of an object through a technological interaction can give us a feeling that it is somehow more natural due to the intimacy of its use (which therefore starts to diminish the feeling of its being a technology under more common definitions). The question then becomes: if writing is as much a technology as a computer or car then how can it come to feel natural in a way that a computer or car doesn't, objects which are so obviously outside of ourselves, particularly when compared to an abstract skill? There is a lot at work in producing such a disconnect between the act of writing and our modern machines, but the primary distinctions are not based around the presence of a material artefact, but instead around frequency of use, particularly during early development, and the length of time that our culture, not just individual beings within it, have refined the interaction under discussion (a point that I'll consider in more detail shortly). For Bolter writing is a technology *despite* being drawn into ourselves, but under the criteria of Incorporation and Domestication it couldn't be a technology unless this was the case.

To approach the issue from the other side, let's consider Plato's critique of writing:

> SOCRATES: They say that there dwelt at Naucratis in Egypt one of the old gods of that country... and the name of the god himself was Theuth. Among his inventions were number and calculation and geometry and astronomy, not to speak of various kinds of draughts and dice, and, above all, writing. The king of the whole country at that time was Thamus... To him came Theuth and exhibited his inventions... [W]hen it came to writing, Theuth declared: "Here is an accomplishment, my lord the king, which will improve both the wisdom and the memory of the Egyptians. I have discovered a sure receipt for memory and wisdom." "Theuth, my paragon of inventors," replied the king, ... ["]you, who are the father of writing, have out of fondness for your offspring attributed to it quite the opposite of its real function. Those who acquire it will cease to exercise their memory and become forgetful, they will rely on writing to bring things to their remembrance by external signs instead of their own internal resources. What you have discovered is a receipt for recollection, not for memory. And as for wisdom, your pupils will have the reputation for it without the reality: they will receive a quantity of information without proper instruction, and in consequence be thought very knowledgeable when they are for the most part quite ignorant. And because they are filled with the conceit of wisdom instead of real wisdom they will be a burden to society." (96)

To cultures like our own, where literacy has become the default, Plato's critique can sound unfathomably flawed. When we consider the social structures, pedagogy, and development of thought that writing allows, it is clear that it is not an inhibitor of education, though it does, of course, impact upon the ways in which we think. Ong suggests that Plato's critique stems not from the revelation of some truth about the inherent unnaturalness of writing, but instead from the act's specific deployment in the Greek society of the time:[38]

---

[38] For further discussion of the emergence and effect of writing in Ancient Greece see Eric Havelock's *Preface to Plato*.

> Plato was thinking of writing as an external, alien technology, as many people today think of the computer. Because we have by today so deeply interiorized writing, made it so much a part of ourselves, as Plato's age had not yet made it fully part of itself, we find it difficult to consider writing to be a technology as we commonly assume printing and the computer to be. (81)

Plato's critique, when we consider it closely, is not wrong; many aspects of specifically Ancient Greek thought have withered under the influence of the move to the almost total ubiquity of literacy in Western intellectual life. But this doesn't mean that literacy has not been naturalised. Part of its power, as with any technology, is not that it exists, nor that it is available (it's a hard-line determinist who suggests that the mere availability of a technology shapes our cultures and ourselves), but instead that it has been adopted so extensively and over such a duration that it underpins the fabric of our society, resulting in Bolter being able to assert that we have managed to trouble some Rubicon between the "natural" and the "merely artificial." Hammers and knives are just as much technologies as literacy, but they affect fewer aspects of our experience when we put away their accompanying tools (which raises the interesting point, to which we must return, if writing is a technology is a pen? And, if so, then if a hammer is a technology, is the art of hammering?).

When we are faced with considering whether interacting with a computer is natural or technological (or merely devicive) we often look to its closest natural (really naturalised) corollary: writing. The computer seems so alienated in the comparison that we assert its technological (unnatural) status without question – technologies, it would seem, are the complex ways in which we (generally) improve upon the efficiency of tasks identified during more "natural" undertakings (e.g., cars for walking, televisions for attending an event, telephones for travel, etc.). But we make a mistake in comparing computing to writing and marking their differences as a sign that one is a technology and the other is a natural, or more natural, process. The difference, in terms of mechanisms of adoption and broad classes of interaction, is solely durational – writing is an example of a technology that has ceased to feel technological following prolonged use far beyond the bounds of a single human life, whereas digital computing is only approaching its third human generation. Literacy can and should be defined as a technology for all skilful users, and we better understand it and its relation to computing (and a pen's relation to computers) by doing so. Let's consider it under the four criteria:

- E – Literacy Extends the capacity of our minds and working memory. It also Extends the mouth and the ears in somewhat abstracted McLuhanist terms.
- C – Another quotation from Bolter sums up the Communal aspect of literacy: it is

  a technology for collective memory, for preserving and passing on human experience. The art of writing may not be as immediately practical as techniques of agriculture or textile manufacture, but it obviously enhances the human capacity for social organization – by providing a culture with fixed laws, with history, and with literary tradition. (33)

  Writing only has use in a Community; the reason of its existence is to pass signs between people separated by time or space, or under pressures of silence.

- I – The quotation from Bolter that triggered this discussion also makes the Incorporation of literacy clear. It has become so deeply Incorporated, so ready-to-hand, so invisible in use that we very rarely consider what occurs when we read or write; it seems as natural as speaking to many.
- D – Literacy has altered our minds in a variety of ways, not least allowing us to approximate an external memory allowing for more complex thought and storage of that thought, giving rise to the notions of legacy and formal culture. Donald also states the more material consequences on the user:

  There is no equivalent in a preliterate mind to the circuits that hold the complex neural components of a reading vocabulary or the elaborate procedural habits of formal thinking. These are unnatural. They have to be hammered in by decades of intensive schooling, which change the functional uses of certain brain circuits and rewire the functional architecture of thought (*So Rare* 302).[39]

To Plato writing was a *device* that devalued what came before it. As it trended towards a technology for more and more users, however,

---

[39] See also discussions of neural plasticity in response to reading throughout Stanislas Dehaene's *Reading in the Brain* and Maryanne Wolf's *Proust and the Squid*.

writing was later able to fully shine, something Plato, despite his own literacy, would never get to see. He was able to set writing up as an inefficient device by comparing it to the "natural" analytic mode of thought cultivated by the Greeks, but, as with comparing computing to "natural writing," might this not hide the fact that analytic thought too was somehow technological?

I would like to pause here only to open a floodgate: I do believe that analytic thought can be seen as a technology, but I also believe that dance, language, the use of fire, and a great polyphony of other things typically thought of as non-technological, or natural, have existed as technologies over the course of human history because they abide equally well by the four criteria of technological interactions that have been set up here. They don't, however, tend to fall under the common-sense definition that relies on a physical artefact that we can point at and label as technological irrespective of the nature of the actual encounter by a discrete user.[40] Such a drawing-in of non-physical equipment also subsumes the philosophy of "technique" – a technique (writing, driving, computing) becomes a different technology to its accompanying device (pen, car, computer), a true technology all of its own.

In many ways what I am proposing returns to the etymology of "technology" identified by Heidegger: "The word stems from the Greek. *Technikon* means that which belongs to *technē*... [*T*]*echnē* is the name not only for activities and skills of the craftsman but also for the arts of the mind and the fine arts" ("The Question" 318). The definition that I have established throughout this chapter is sympathetic to this originary etymology: a universal understanding of technology as a *type of experience* need not apply solely to interactions with material artefacts, nor to

---

[40] There are many precursors to this idea. Foucault's four definitions of technology in "Technologies of the Self," for instance, are expansive in what they allow to be considered as technological independent of the use of artefacts. Lev Vygotsky, in *Mind in Society: The Development of Higher Psychological Processes* and "The Instrumental Method in Psychology," also argued for the existence of "psychological tools" such as language, writing, mathematics, drawing, mapping, etc. which overlap with my notion of atypical technologies, functioning as intangible equipment. That said, the argument here is that the definition outlined in this chapter is potentially more capacious than that of either Foucault or Vygotsky, as well as offering a more particular reasoning for why such things should be considered as being technological under certain conditions.

the results of actions that take the form of external representations. This is again best illustrated by an unintuitive example:

Dance[41]

- E – Dance Extends an individual's ability to communicate, to express herself.
- C – Dance functions as a technology in a Community which is aware of (most of) the implications, the meaning and nuance, of the movements of the dancer. Part of the Extension described in the point above is the need to communicate ideas which are either best expressed non-verbally, or cannot be expressed verbally, or to draw a group together via their standardised or recognised kinetic expression.
- I – Dance is Incorporated by expert users, and this is evidenced by actions which could not be consciously contemplated if they are to be strung together fluidly. This ability begins by the dancer making themselves supremely aware of the shape of their body in space during different motions, typically via pedagogic correction, through mirrors, or even via audience reaction. This heightened consciousness is then rehearsed and rehearsed until it disappears, for the most part, though it is still able to be interrupted by unexpected success or failure. The result is fluid expressive motion, where the dancer thinks about what it is that they want to convey and their general motions, rather focussing on the minutiae of the micro- and macro-movements required.
- D – The dancer's body is changed by the practice of dancing, growing leaner, more flexible. In more casual and undemanding practice, where gross physical alteration is far less, a dancer will still have her mental image of herself altered, a heightened awareness of her physical capabilities, and a way of thinking that is not coded linguistically, but visually, physically, kinetically – a sense of new potentials becomes engrained.

But there are at least three technologies at work in dancing. The art of dancing is described above, but the dancer could also make a particular dance (equipment for meaning) into a technology, and even experience her own body technologically, as equipment for dancing (we will return to this last idea very shortly).

---

[41] For more on the phenomenological experience of dance see Danielle Suzanne Vezina's *Phenomenology and Dance*.

If we shift the requirement of defining technology from being artefact-centred to *equipment*-centred, then the four criteria begin to hold in such unexpected and overlapping encounters. This allows for what I would argue is a more accurate view of how humans experience their world: we encounter a host of things in use, and focus on specific aspects of their and our being through deployment – practices, objects of the mind, and ways of thinking are no exception.

The power of this conception of technology lies in its allowing us to see how technological use, that is, a particular flavour of experiencing the world (and perhaps our most significant mode), repeatedly plays out throughout human experience.

Taylor, in his exploration of Tasmanian Aboriginal tool use in *The Artificial Ape*, offers an illustrative case (33–54). Before they were wiped out by European settlers, nomadic bands of Tasmanian Aboriginals survived without clothes, may not have been able to make fire (instead carrying it with them in "fire logs" after discovering its natural occurrence), and rarely built roofed domiciles. This led to a society in which there was very little private property, nor much in the way of social hierarchy. Taylor argues that, contrary to prior anthropological opinion, these people were not somehow "backward" or tragic, but instead may have been perfectly adapted to their environment, minimising risk as best they could, and only keeping what was absolutely expedient. Commentators, both from the time of the European encounter and reporting since, who had compared the tools available to the Tasmanians to those used by chimpanzees missed the point says Taylor:

> The more we look at the Tasmanian Aboriginal toolkit, the less the parallel with the tools of chimpanzees (legitimately enough made on formal grounds) makes sense. It is not just that the humans had more things, because with only two dozen items, it was not that many more. It is that their technology was not an add-on, an optional extra. It was essential and embedded. Chimps can live without tools. Humans cannot. (*Artificial* 52)

The Tasmanians of this period were as technologically minded as any other human society because they did not just elect to use tools, but instead used precisely what was needed to thrive in their environment; they could not live without them. That their artefacts are amongst the simplest collection in the *Homo* archaeological record did not make that equipment any less vital or suitable to their purposes – with so few artefacts it is likely that each was expertly used; the Tasmanians may have had the highest ratio of technologies to devices in history.

By reconsidering what we mean by the term "technology" we might be able to come to the kind of nuanced reading of cultural difference that Taylor outlines more often, rather than seeing "advanced" or "developed" societies as having so progressed that all others, at least (but rarely just) in terms of technology, are now just waiting to catch up. Technology is at the heart of all human experience of the world; the only thing that marks out a developed society's technologies are the material specificities of, for example, horse vs. car or the differing complexity of the tool's production in, for example, book vs. e-reader. It is not even that such artefacts take longer to master, but that basic access (and certainly manufacture) is often more complicated than that of more simple tools. Most anyone can throw a spear to some degree, but few can hunt in such a fashion; it may take longer to acquire the basic skills of using a car, e-reader, or computer, but does something that we would recognise as expert use, as mastery, really take significantly more dedication than that of the hunter to her chosen equipment? More importantly, does a society which sees many people wake up, hunt, skin, and prepare food, make tools, make weapons, and engage in a participatory (rather than passive) creative culture really seem less technologically minded, if we consider technology as rooted in the uniquely rich human experience of the world through equipment, than a society where many people wake, drive to work, input at a computer, drive home, heat food, and watch television? The artefacts in the latter experience are certainly of a higher complexity in terms of production and, sometimes, in attempting initial use, but are the interactions really stronger in their impact on lived experience, or their use more skilful? Domestication, Incorporation, and an equipment's embeddedness in a Community, I would argue, are far greater measures of the strength of a technology's effects and centrality to life than the complexity of an artefact. The forms that we tend to actually encounter are also typically *not* harder to use than a bicycle or a spear or a violin. It's a strange hubris to say that when you microwave a meal you are engaged in something more complex than a concert performance 10 years or more of training in the making; that when you struggle to set the clock on your DVD player you might wish for the simplicity of older or more "primitive" societies where you simply had to craft and deploy the tools of the hunt and the butchery of its outcome.

Computing has had a profound effect on society, changing it dramatically. We might want to say that this marks out developed cultures: the profundity of the change to life that the extensive use of technology enacts. But again this implies a pre-technological norm that we've deviated from – hunting and farming, knives and hammers, each changed human culture far more profoundly, and were deployed far more extensively, than computing will be for a considerable time to come (that we can even imagine it coming close is indeed a marvel).

To conclude, let's consider one final piece of equipment that would traditionally be considered well outside of the realm of the technological; let's look at whether the body itself can be encountered as a technology. Consider the schematics from Dotov et al. reproduced above, where an interconnected system of black dots represent elements of the body as an incorporated assemblage into which the tool can be introduced during skilled use: we have to *learn* how to manipulate our bodies like this. Clark makes this point explicitly: "The human infant must learn (by self-exploration) which neural commands bring about which bodily effects and must then practice until skilled enough to issue those commands without conscious effort. This process has been dubbed 'body babbling'[42]... and continues until the infant body becomes transparent equipment" (*Supersizing* 34–35). In our infancy we are a collection of parts held together by skin, our brains are a mess of connections which need to be whittled into some useful shape, and our use of the equipment, not that we have, but that we *are*, is less than skilful. Over time we rehearse the manipulation of ourselves into Incorporated unconscious use, and do so by taking ample cues from our Communal surrounds. Such skilful use certainly Extends the range of options available to the body, and the act of Domesticating ourselves into adulthood is the most profound enacting of the criterion that we will ever pass through.

We can also approach the issue from the other direction and see whether bodies can be brought out of technological use and made into devices, moved, in Heidegger's terms, from readiness-to- to unreadiness-to- or presentness-at-hand. Activities like dance, or sport, certainly mark a return of our bodily equipment to consciousness as we learn the limits of our abilities. Psychiatry similarly aims to make aspects of our psychology present to consciousness, and brainwave imaging has allowed people's hidden neurological activity to become available, in a limited way, to their contemplation, allowing them to try and manipulate the real-time images onscreen by relaxing/tensing, etc.[43] At most

---

[42] See, for example, Phillippe Rochat's review "Self-Perception and Action in Infancy."

[43] See, for example, Scheinost et al. "Orbitofrontal Cortex Neurofeedback Produces Lasting Changes in Contamination Anxiety and Resting-State Connectivity."

times we encounter our body schema as invisibly working equipment for our dealings with the world, but we are, perhaps, the only species which can consciously reflect on aspects of such an engagement, from metacognitive analysis to a free-diver's hyper-awareness of her breath, from yogic meditation to athletic refining of form. For these engagements a specific configuration of aspects of the body, the encounter with what Gallagher would call "body image," where the schema is tailored temporarily to a particular use, is made available to consciousness and can then be re-rehearsed back into unconsciousness, altering the nature of the global imagination of ourselves. When the body is made into equipment for use, the same experience becomes manifest; it is not the whole physical form that is encountered, but instead merely those required aspects which must be rehearsed and then trained into invisible automaticity. We do not encounter every aspect of ourselves as technological, but we can bring elements of ourselves out and bring them to consciousness as machines-in-order-to-do-X.[44] We can, and do, make artefacts of ourselves and skilfully deploy the encountered elements in much the same way as we might a hammer, racquet, or car.

An argument against the theory deployed here might now stem from the observation that if the body schema, indeed the body itself, can be thought of as a technology, alongside a whole host of more or less exotic things, then what *isn't* a technology? But, of course, I'm not suggesting that everything is a technology, nor that our every interaction is a technological interaction, only that very many things *can* be. Not everyone is an athlete, not everyone learns a new language, sees a psychiatrist, reads a book, drives a car, or skilfully wields a hammer. The vast majority of the average person's interactions with objects in the world are devicive, but near everything is also theoretically available to be technologised. This is the curiously human outlook, what may separate us from other all other animals: our ravenous search for opportunity, to apprehend every aspect of an environment, including ourselves, as exploitable when needed. For Heidegger, in "The Question Concerning Technology," this "enframing" is the reductive modern stance, but I would argue, instead, that it is ancient, powerful, and necessary, and we will go on to look more closely at the importance and inevitability of reduction in the following chapters. The awareness of our limits, physically and intellectually, is part of our basic cognitive agenda. We constantly check and recheck those limits, establishing the shape and reach of ourselves, and

---

[44] We'll look further at the issues of parts and wholes in technological use in the next chapter.

try and work around any resistances that impede the abilities inherent in the schemas and images that we produce. This, I would argue, is the true scope of technological use.

2.6 Coda

Let's return to our original common-sense definition: technology enables, technology is a uniquely human thing. I hope that this definition now appears deficient as a basis for any nuanced discussion of the particular phenomenological experience of equipment which is at the heart of human experience. When we say that technology enables us to do things that we were unable to do before then this is perfectly correct, but should every object that allows for new abilities be called a technology? What about reaching for one's car keys, lost at the bottom of a drain, with a length of bent coat hanger? Is this a technology, a technological interaction? The common-sense definition would say yes, even though it seems far from common-sense to see the clumsy swipes with a makeshift key-collecting hook as being of the same order of experience as the expertly used computer or car. As Wallace, Taylor, Heidegger, Maravita and Iriki, Dotov et al., Wilson, Donald, and Clark all assert in their various ways: skilled use changes the order of equipment that we encounter.

The definition outlined here has drastic implications for a number of the artefacts which we would readily describe as being technologies as we are forced to ask what I have tried to position as a phenomenological question: who are they technologies for? Let's take a final example of a passenger plane (the object, not the act of flying). The common-sense definition states that a plane is unequivocally a technology – humans use planes to do something that they previously could not. But when we deploy our revised definition the question arises as to when we should say that a particular user's interaction with the artefact should be considered as technological.

Overleaf there are three users' interactions broken down into the four criteria:

The definition established here, when deployed, overhauls our consideration of mass aeronautic travel as a technology; planes are a great example of an item that most of us *do not* encounter in a technological interaction. We are conscious of it (i.e., it is not Incorporated), well outside of the Community of those who actually use it in a skilful sense, and we are largely not Domesticated by its existence. It does Extend our ability to travel, and if Extension were our only criterion for a technology, as it is in the common-sense definition, then it would be met. But it shouldn't be – this enables us to distinguish the passenger's

	PILOT	ENGINEER	PASSENGER
EXTENSION	The pilot is able to achieve an activity – that is, flying, or travelling great distances, or movinat they could not have achieved before.	Not applicable. The *plane*, as experienced during work undertaken, does not extend the abilities of the mechanic; they work on the plane with skill-extending tools.	The plane allows for a new, though relatively uncontrolled ability: the ability to travel at speed through the air.
COMMUNALITY	The pilot has been through flight school, experiences cultural pressures at various levels (global, national, professional, fraternal) to interact with the plane in a certain way.	The mechanic experiences a variety of cultural pressures to interact with the plane in a certain way.	There are specific cultural pressures surrounding the act of flying which constrain, codify, and standardise the experience.
INCORPORATION	The pilot's experience of the plane as artefact will principally come from interactions with its external boundaries (an awareness of its size for taxiing and turning) and the cockpit. The pilot can be situated as a *component* within this assemblage, and this is also a form of incorporation.[45] Pilots are able to perform activities functionally invisible to themselves: all the micro adjustments of pitch and yaw done by "feel,"[46] every unfocused reach for a dial to make a correction, or the automatic elements of pre-flight checks.	Whilst the mechanic is primarily incorporating her tools during skilful work, she might also have a sense of the feel of the plane, very different to the pilot's, but again born of repetitive engagement. The intuitive checks and adjustments of the pre- and post-flight equipment, and a sense of the whole operating smoothly might well be considered a form of incorporation.	Not applicable. Flying on a plane is a supremely conscious experience, and there is no way to incorporate the artefact that a passenger experiences. The passenger has no access to use the thing that they encounter.
DOMESTICATION	The pilot's body schema has an increased range of abilities available to it. Hand/eye coordination will have been elevated; a feel for the plane will have emerged.	Not applicable. The plane, as experienced during the work undertaken, does not have a domesticating effect on the mechanic.	Not applicable. The plane itself is unlikely to have a lasting impact upon the passenger.

---

[45] See, for instance, Edwin Hutchins discussion of airline pilots and their equipment in "How a Cockpit Remembers Its Speeds."

[46] By feel I mean what, to the pilot, seems an ineffable sense of the whole: they make adjustments that feel right for the assemblage in which they situate themselves, and, far more often than not, they are the right adjustments. This stems from a deep skilful incorporation born of repetition, similar in effect to a dancer or gymnast's almost uncanny awareness of the shape and position of their physical form.

interactions with the plane as an artefact from those of the skilled pilot and the skilled engineer, and also to distinguish between the order of encountered objects that includes expertly used hammers and mobile phones, and the order of encountered objects that includes passively sitting in a metal carrier.

The definition presented above is intended as a challenge to the discourse that sees new technologies as being somehow unnatural – technology, here, is the most natural thing about us. But by so naturalising technology, or by saying that computing and dance are of the same order, does this theory potentially do damage? Does it weaken the important task of questioning the deployment of new technologies, including widespread e-reading? I would argue not; it shifts the fight from worries about technology as an abstract and general concept, to worries that are more specific – we should be vigilant as to how new artefacts and new ways of thinking are deployed, by who, and to what ends, but that task is not best supported by a misunderstanding of our nature as skilful users of new opportunities. It is, in short, a move towards specificity, not a neutering of the debate. If we understand dance and computing as related – though clearly fundamentally different in their praxis, function, and effects – then we might better avoid irrational fears based on false binaries, and instead start to home in on any real problems with our use of certain equipment.

Once a reliable technological encounter has been introduced we find it hard to challenge. Expert use is valuable, technological systems are hard to set up, and it is difficult and time-consuming to rehearse devices into invisibility; there are, in short, many good reasons to be resistant to change, maybe even as a survival trait. As such, replacing what appears to be a fundamental technology (bound books) with what appears to be a hopelessly visible device (e-readers), perhaps unsurprisingly can feel like a weak move, bereft. But over the course of writing this book the reports from popular media have started to change as more and more people have begun to use tablets as their standard media device. The worries about e-reading persist, and yet increasingly the devices melt away:

> there's something seriously different about [the iPad]...That difference can be summarized in two words: It disappears...Instead of living inside a box with a URL bar and a bunch of buttons alongside other boxes and applications, content takes over the device...You're not just looking at [the internet] through a browser, you're holding [the internet] in your hands. (Tweney)

If the benefits are worth it, and we can get used to the tools, then the technologising process is well under way; it's what we do. It is to this idea of "getting used" to new equipment and the specific nature of our encounters to which I will now turn.

3
All Is One but Not for All: Technology as an Object Encountered in the World

In the first two chapters I was interested in exploring the language that we use to talk about technology: what that language reveals about our fears, how such language might lead us astray when we talk about what technology does, but also how we might deploy a taxonomy of effects and affects in order to better understand our expert encounters with equipment. This chapter and the next focus, instead, on exploring our physical and conceptual experiences with technology, how we meet it, and what it does. For this chapter I want to consider what it is like to encounter a physical artefact as equipment, as something to be used, and the ways in which that encounter is structured by our prior experience. I want to explore in particular how those artefacts that we use most closely, technologically, are both objects and shapers of cognition and perception, how they are both moulded in and come to mould our minds.

I'll begin by discussing how we encounter artefacts as "whole-composites" or "gestalts," as changeable collections of elements that we typically experience as single things.[1] Drawing on Husserl and Merleau-Ponty's explorations of objects' "adumbrations" and "horizons," I want to suggest that our encounter with any object, and in particular our tangible technologies, is never complete; some element must always escape our comprehension. This continual *escapability* or illusiveness of an object's aspects is vital for ready-to-hand incorporated use: we cannot keep everything in mind when we are being skilful; we reduce the artefact to the absolute minimum of itself that is required for a successful interaction. But, as I will go on to argue, this reduction also allows for

---

[1] That this is true of all objects will become clear, but our modern complex artefacts offer a particularly compelling example, and it is during expert use that the effects of such reduction are most pronounced.

the artefact's breaking back into our consciousness as some ignored aspect returns the whole to our attention. Objects always retain the capacity to surprise us, and all the more as our expert use reduces them to a minimum of aspects. I will favour a language of escape as artefacts, and objects more broadly, have escapological tendencies – perception and description attempt to place boxes around things, to bend them to their fit; they rarely stay there, and never neatly, they have a tendency to bolt. Encountering an artefact as a technology, then, is a strange thing: a set of effects which cause us to experience an object as a member of a phenomenological class (technology) that relies on our never encountering that object with any richness – a reliable technological use is a balancing act between a minimally effective set of encountered features and the threat of surprise from some hidden aspect. The development of expertise involves cultural support and individual trial and error in determining what this minimum set might be; as a prelude example we might listen for the ring of the hammer as it strikes, and sense the torque to left or right, in order to prime the next blow whilst wholly ignoring the grain of the wooden handle that we hold and the gradually weakening glue that unites haft and head.

That objects always appear to us in some reduced and unfixed way enables me to talk about three things: principally, for this chapter, to deepen our understanding of how we approach artefacts, particularly those that are new to us; then to give a phenomenological, and later ontological basis for why amateurs and experts, quite literally, encounter different objects during the move from device to technology; and, finally, to account for the moments of surprise, glitch, and rupture which will always drag an object back from being a technology and into more devicive use. These elements each stem from making the same point twice: firstly saying that we encounter artefacts as reduced gestalts and deploy metaphors in our attempts to get used to objects, transferring knowledge from one domain to another and thereby altering the composites that we encounter. And then, in Chapter 4, using these ideas of approximation and combination to establish the fundamental ontology that underpins the conclusion of this work, I will explore an object-oriented approach drawing on the work of Graham Harman. I intend to show why Harman's object-oriented ontology (OOO) might be of use to the philosophy of technology and how it gives an explanation of technology's continual capacity to evade and to surprise us. Contra Harman, however, I want to suggest that the process of technologising outlined in Chapter 2 is a development of our access to the artefact as-it-is, and this will enable us to step beyond the human user to consider artefacts from a posthuman vantage, on their own terms, as embodiments of knowledge.

Before we depart from the human, however, this chapter will conclude with a discussion of Don Ihde's postphenomenological approach to technology and the findings of contemporary Cognitive Science (particularly Anthony Chemero's *Radical Embodied Cognitive Science*) that see cognition as a rich interplay between brain, body, tool, and environment. This prior research will help in demonstrating the ways in which the embodiment of user and artefact contribute to the specific gestalt object that we meet.

This chapter is intensely focussed on human encounters, human use, but the ideas established here represent another stage along a trajectory away from over-simplification. In Chapter 1 I was interested in the misunderstanding of how important technology is to humanity; in Chapter 2 I wanted to do justice to that importance with a nuanced description of our use; and in this chapter I want to more fully explore those encounters, the things that come to mind and the things that escape us. The final chapters, however, will take us to our last remove, going beyond the individual human user in favour of paying attention, as Husserl intended, to the things themselves, how they come to be and what kinds of information their bodies hold independent of all contact. But first we need to establish how we conceive of them at all.

3.1 Encountering elements

In Chapter 1 we saw a thread of folk-phenomenological reports of e-reading equipment not feeling right, being unnatural, partly due to the sense that the text is somehow placed *behind* the object, that there is an additional layer of visceral insulation that must be fought through before it can be accessed, an insulating layer that isn't present with the codex. We can assume, therefore, that at least some users experience disruptive formal differences between these artefacts, between the e-reader and the printed book. But I'm also interested, here, in how different users, or the same user at different times, can encounter similar formal differences within the *same* object. This needs to be accounted for as it is a part of the distinction between amateur and expert use that my definition of technology rests upon. Put simply: when you get better at using something it's not just you that changes.

When I owned an old cathode ray television, a big black box behind a thick square screen, the object that I encountered was made of glass and plastic. It had a tired-looking remote, each of the rubber buttons were worn smooth and blank and I knew exactly what each one did and how hard I needed to press them in order to get a response. The television and its remote were simple; I encountered a few components in three materials: plastic, glass, and rubber. One day my television broke and a friend insisted that he could probably fix it; I think that we both just wanted an excuse to open up that big black box. We set out a thick towel, unplugged the TV, and hefted it screen-side-down onto the floor. We undid its screws. These were previously un-encountered spirals of metal, something new – the TV, on its face and with these new elements exposed, was becoming strange. We lifted away the casing and saw: brown circuit boards, wires, stacked metal components I still can't name. My television was a city. Somewhere in the heart of all that, there was a cylinder. My friend didn't have any idea how to fix my television, and he also had no idea, neither did I, that the cylinder was a capacitor with enough charge left in it to kill us both and maybe start a decent fire. Luckily we put the case back on, baffled by what we'd seen.

No large television has ever looked the same to me again; I now know some of what's hiding beneath the surface, like a junior doctor after her first explorations of a cadaver. The object that I encounter has fundamentally changed in some way, and part of what we need to deal with is whether it's just the object in my mind that has altered, some mental representation that has expanded, or whether this is a real new object on its own terms (as we'll go on to discuss in Chapter 4). Even if it is just the representation that alters, then it certainly has real world effects.

Derrida, describing the experience of another (literal and figurative) black box, notes that

> even if people know how to use [a computer] up to a point, they rarely know, intuitively and without thinking – at any rate, I don't know – how the internal demon of the apparatus operates. What rules it obeys. This secret with no mystery frequently marks our dependence in relation to many instruments of modern technology. (*Paper* 23)

That beautiful phrase, the secret with no mystery, describes the experience of complex technologies, the typically un-encountered aspects of our artefacts that are theoretically knowable, but only if we take the time to investigate them and to learn. For a computer scientist (or television engineer) there is, in this sense, little mystery within the box. I want to go on to suggest, however, that secrets, and sometimes perpetually mysterious ones, might characterise aspects of all objects, that there will always be some secret of a thing which withdraws from our experience.

But for now it is enough to note that, for the average user, and particularly with modern complex artefacts, we tend to experience a very partial selection of the whole.

Any discussion of physical technology, then, must ask "what *is* the artefact that we're discussing?" An artefact is rarely one thing: is the mobile phone the technology or is it the circuit board, the screen, the 3G connection, or the Wi-Fi receiver? Is the axe the technology or is it the flint head, the shaped and wrapped wooden haft, or the hand-made twine that binds them? We might intuitively say that it is the sum total, that the artefact is the item of equipment at the nexus of a set of other equipment; each artefact, in this view, is a technological system of sorts. But what we experience *isn't* a slew of components and wider systems – we always encounter, in smooth technological use, an object that feels whole.[2] The technology, then, (and this abides with the view of technology outlined in Chapter 2) is the object encountered in use, the one thing. The computer is not a technology when encountered as a sum of complex parts; the computer is a technology when the keyboard, mouse, and screen (and each of their components in turn) melt away during expert use.

A technological artefact might be constructed from a range of artefacts that in themselves may or (more typically) may not be technologies to the particular user (or to others); we saw a little of this in the different experiences of the plane to the pilot, engineer, and passenger. When we read reports of physical books as being more "natural," as we've seen it's not that they're not technological, but that their encountered equipmental reality makes an intuitive sense. The act of reading on a digital reading device, however, is built around a set of Derridean secrets: unintelligible, non-technological processes that allow activities that *can* be technological to run on top of them.

This gives us a strange structure for multi-part technologies: the technological object isn't what the artefact is in some "truthful" totality, but is, instead, an object in its own right and encountered as such. It is a "whole-composite" (the perception of the complex as singular) with its own effects distinct from the object of some scientifically described whole, even as the particulars of that real object constantly threaten to intrude.

---

[2] We can nuance this further: the harder it is to cleanly separate parts of an object with our hands, the more we typically consider them to be elemental – a cup's handle is fundamental, a lid is not; a book's cover is, but a sticker is not; a seamlessly glued table seems a more whole object than one obviously screwed together. An object's physical boundaries for us are therefore most often defined in relation to our bodily experience of them.

Another way of describing the whole-composite form of the encountered artefact would be as a "gestalt." In *Metaphors We Live By*, the cognitive linguist George Lakoff and philosopher Mark Johnson use the term to describe the combinations of elements that "recur together over and over in action after action as we go through our daily lives... [where] the complex of properties occurring together is more basic to our experience than their separate occurrence" (71). For Lakoff and Johnson, a gestalt is a convention, a learned way of encountering a complex of elements as a single more simple or more useful entity that might otherwise be abstract or beyond our comprehension.

The naming operations described in the opening paragraph of Deleuze and Guattari's *A Thousand Plateaus* offers an example that abides by the same principle: "We have assigned clever pseudonyms to prevent recognition. Why have we kept our own names? Out of habit, purely out of habit... Also because it's nice to talk like everybody else, to say the sun rises, when everybody knows it's only a manner of speaking" (3). Names don't refer to singular coherent beings for Deleuze and Guattari; they are convenient signifiers to describe the polyphonic mass occurring in and through a body, as convenient as saying "the sun rises" without having to think of the optical illusion created by planetary orbits. We encounter ourselves as gestalts continually: as approximations instead of the polyphonous selves described by Deleuze and Guattari,[3] and as a simplistic skin-boundaried (and often internally empty) embodiment. In *Descartes' Error* the neuroscientist Antonio Damasio, for instance, identifies "somatic markers," the relative sameness of signals from the body that we experience each day, as the defining reason for fixity in our sense of self despite a much more complex reality (165–203). But sunrises, neatly bloodless and organless conceptions of the body, and coherent senses of multiple selves aren't simply lies – the gestalts that we encounter always have real effects on their own terms, not just as approximations. For Damasio, our sense of self is possible not simply due to our not noticing, our not paying attention, but due to our *paying attention* to a very particular thing as itself, that is, the near-as-damn-it continuity of day-to-day embodied experience. Similarly, Deleuze and Guattari's pseudonymous selves certainly have a real effect on how people conceive of their ability to form an identity.

For our purposes here, it is important to note that our gestalt encounters with artefacts are similarly effective:

---

[3] "The two of us wrote *Anti-Oedipus* together. Since each of us was several, there was already quite a crowd" (*A Thousand Plateaus* 3).

Previously, as we were cleaning [motorcycle engine] parts, I had held one of [the] valves in my hand and examined it naïvely, but had not noticed the mushrooming [flaring at the top of the valve head causing friction]. Now I saw it. Countless times since that day, a more experienced mechanic has pointed out to me something that was right in front of my face, but which I lacked the knowledge to see. It is an uncanny experience; the raw sensual data reaching my eye before and after are the same, but without the pertinent framework of meaning, the features in question are invisible. Once they have been pointed out, it seems impossible that I should not have seen them before. (Crawford 91)

Matthew Crawford, a philosopher by training and now a professional motorbike mechanic (by more rigorous training), here describes the experience of encountering the same artefact in two different ways: uninformed and informed; amateurly and through the eyes of an expert; as a smaller and larger gestalt of aspects. And this readily demonstrates the ways in which each encounter enables different activities in the world; the first encounter, where the issue isn't seen, allows the part to be reinstalled and to fail "mysteriously." The second encounter allows the part to be fixed and any issues prevented. In this way, it is always the gestalt that determines what actions can be conceived of as being enabled by an implement, not some "realistic" totality of the artefact's true strictures.

The gestalt of a computer, Derrida's secretive black box, is both idiosyncratic and malleable: I can always add to what the word "computer" means to me. I encounter the computer, or the e-reader, or whatever, as a delimited, but non-intersubjective thing (even if I say to you "that computer over there") because the world's computer users are unlikely to agree on what exactly it comprises of.[4] To borrow

---

[4] Gestalts'/whole-composites' non-intersubjective nature will increasingly resonate as we move on to the next chapter. The object that I encounter is always the object for me, not for you – you always encounter a different object. This certainly isn't saying that there are no real things, but as I will go on to argue in Chapter 4, we never have access to things as they are even as those real things produce real effects upon us. The ways in which we form personal and un-shareable gestalts discussed in this chapter gives us a way-in to the more radical implications explored next, including that intersubjective validity might traditionally be a component of knowledge, but it need not be a mark of access to reality.

a term from Husserl, the "adumbrations" of the specific computer alter as we engage with it in different ways, and this leads to an altering of what we intend by the word. An adumbration is the reduced way that an object manifests itself in perception to a viewer with a specific vantage, that is, as an adumbration of the real whole – "[o]f necessity a physical thing can be given only 'one-sidedly'" (Husserl, *Ideas* 82).[5] As we'll return to in Chapter 4, Husserl maintained that the encountered object retains a fixity independent of its various adumbrations (e.g., we know that the television viewed from the side is the same television when viewed from the front).[6] When our vantages enable us to include more and more components into our gestalt, however, there is a real phenomenological distinction that needs to be taken into consideration – there is a change in *richness* in an object explored, what Husserl calls a move to an "intuitional fullness."[7] For Husserl this is a fleshing out of the object as-it-is, but, as I will continue to explore here, this need not be the case; the alteration of a gestalt isn't always about accuracy; there is a wide variety of provocations to changing what we encounter. As Crawford demonstrates above, for instance, such a change can be provoked from outside of ourselves, beyond our own experience and use; we can take on and test out another's vantage, and this too can be vital for the development of skilful use.

We might also deploy another Husserlian term in this regard: "horizon." An object's intentional horizon is what gives a perception of an object its fixity and fullness; just beyond what we cannot currently perceive there are features that we "anticipate" – "the perception has horizons made up of other possibilities of perception, as perceptions that we could have, if we actively directed the course of perception otherwise: if, for example, we...were to step forward or to one side, and so forth" (*Cartesian Meditations* 44). If I walk around my television then I know that it has a back and sides; if I open it up then I know that it has circuit boards and screws and wiring. In this way anticipation is an

---

[5] As Don Ihde describes the experience in his work on postphenomenology: "A complete phenomenon...has both a manifest profile and a latent sense...I do not see the world without 'thickness' nor do I see it as a mere facade. What appears does so as a play of presence and a specific absence-within-presence" (*Experimental* 41).

[6] See Husserl, *Ideas* 82–85.

[7] See Husserl, *Logical Investigations* 246–248.

embodied experience[8] – if I move my body in this way then I know what the effect will be, and those "missing" aspects that I know would be revealed are a part of my perception for Husserl; they are "co-present," just waiting to be brought to light.[9] My understanding of any object as what-it-is is dependent on this fact: "[Husserl] is not merely arguing that every perception of an object must necessarily include more than that which is intuitively present; in order to see something *as* [e.g.,] a tree, we will have to transcend the profile that is intuitively given and unthematically co-intend the absent profiles of the tree" (Zahavi 96). All objects appear, therefore, against an interplay between perceived presence and anticipated absence, and such anticipations must be conditioned by prior experience, the extent of which will affect the reliability and accuracy of our anticipations. That an iPad will have a reverse side, that it will have weight, that it will feel smooth, are all reasonable expectations from my prior experience of encountering objects in the world, but its relative thinness, lightness, heaviness, or smoothness can be better anticipated the more that I engage with it or objects similar to it.

But this sense of the co-presentness of hidden aspects, I want to suggest, is something hard won, never fixed, and never finished (this marks the real distinction between Husserl's faith in moving towards an "intuitional fullness" and my continually flexible and necessarily

---

[8] Perception is always embodied; we cannot ever perceive more than our bodies, with or without augmenting artefacts, can provide data about. This may seem an obvious point, but it is one that is typically forgotten in daily experience, particularly in its anthropocentrism. Dogs see fewer colours than we do (David "Color and Acuity Differences Between Dogs and Humans"); bees can perceive ultraviolet light (Peitsch et al. "The Spectral Input Systems of Hymenopteran Insects"); and birds may even witness quantum entanglement (Grossman "In the Blink of Bird's Eye, a Model for Quantum Navigation") – which of these creatures sees the world as it is? We struggle to even comprehend how these organisms encounter their environment, and objects within it, and their experiences reveal the lack in our own. But our bodies are also what makes our brittle attempts at knowing possible.

[9] See Husserl, *Cartesian Meditations* 44–46. Jorge Luis Borges' "Funes the Memorious" also comes to mind, a short story describing the fate of Ireneo Funes, a teenager with a perfect memory of every aspect of the world that he has seen. Funes cannot conceive of how the word "dog" "embraces so many unlike individuals of diverse size and form; it bothered him that the dog at three fourteen (seen from the side) should have the same name as the dog at three fifteen (seen from the front)" (*Labyrinths* 93–94). Funes' perfect memory means that he doesn't unite his past experiences to produce anticipations; every moment of perception is fractured and nameable as something distinct – every moment is an object of its own to Funes, rather than full of objects which persist independent of the vantage taken on them.

imperfect whole-composites). I can always know more and I'll often probably never bother to look; using a technology skilfully, after all, doesn't require us to know all of its secrets. When I say "computer," if I don't know that something is a part of the artefact then I'm not referring to it when I say the word, I'm just setting my own personal boundaries of intention (and therefore interaction) under a word that everyone else is using to do the same thing; I'm saying that the sun rises.

Similarly, the internal workings of my television weren't co-present until I was persuaded to investigate, and now that I live in a world of LCDs, rather than cathode rays, I have a strange co-given anticipation of something that's not there – I expect there to be elements that likely don't exist, but they still have a real effect. I have no idea if an LCD television has a capacitor inside it capable of killing me, but I'd act as if it had – for my gestalt experience of the object it might as well. In this way, experiencing objects as whole-composites with an essential character differs from the dream of access to reality through perception and contemplation that Husserl's phenomenology aspired to. It is because we perceive flexible gestalts with potentially false co-present aspects that have real effects that we do not encounter things "as they are."

To state it plainly: it must be the case that we encounter things as gestalts rather than some more simple direct perception or eventual intuitional fullness because we can always be surprised (a major theme of Chapter 4) and we can alter our perceptions to include (typically more subtle) information that we can base future actions upon (which we'll focus on here). The violin that I encounter is not the same as that met by the concert violinist; my computer isn't the same as for the computer scientist; seeing a Formula One car I don't perceive what the professional driver climbs into. That these are real-ly the same objects is clear from repeatable successful action: the violinist says "draw the bow across the strings and it will make a sound" and she's always right and right for everyone who tries to do so. But that I encounter these objects differently from the experts is equally true, revealed by the real effects of what I encounter in comparison to the real effects of what they encounter, what we can conceive of performing, and our particular anticipations. I don't drive, and I have no idea what the underside of a Formula One car looks like; what is co-present to me looks much like the underside of a Ford Fiesta I saw up on blocks in my teens, and this necessarily structures my actions, intentions, and prehensions.

We must now turn to how we produce the gestalts that we encounter. I will argue that whole-composites are a product of the particular qualities of the real artefact meeting qualities of our enculturated bodies and embodied cognition. I want to think about these elements in terms of the stories that gestalts tell: the stories of our past experiences; the stories we read in the world; the stories told to us by our embodied interactions and by the cultures that we're embedded in. Each of these elements affects our encounter with particular objects that are real on their own terms and yet, as I'll continue to establish, they produce an idiosyncratic composite with its own particular and real effects. By exploring how gestalts are produced, as blends of embodiment, metaphor, affordance, perception, reception, hermeneutics, and heuristics, we can start to understand how malleable and impactful a gestalt might be and therefore better explain the distinctiveness of amateur and expert use. If technology, as I have argued, is about a particular kind of skilful use of equipment, then the transition from device to technology is, at least in part, about shaping a more reliable (if not necessarily accurate) prehension of the equipment that we intend to use.

3.2 Structuring gestalts with metaphor

The essence of metaphor is the "understanding and experiencing [of] one kind of thing in terms of another" (Lakoff & Johnson, *Metaphors* 5). This seeming simplicity belies the immense enabling potential of metaphorisation. Metaphor is central to how we think, it "pervades our normal conceptual system. Because so many of the concepts that are important to us are either abstract or not clearly delineated in our experience (the emotions, ideas, time, etc.), we need to get a grasp on them by means of other concepts that we understand in clearer terms" (115). Lakoff and Johnson see metaphor as drawing on our embodied experience, our enactive progression through day-to-day life in order to comprehend the abstractions that we have grown to rely upon. The perception of time as spatial, directional, is a familiar example: time in the future is up, or ahead of us,[10] the past is behind or below us. The abstraction of time is often best comprehended by our embodied experiences from other simpler and more tangible activities, and the language that we use gives us clues as to how we experience these ideas and further promotes and reinforces them (*up*coming events, what lies *ahead*, etc.).[11]

---

[10] And I've always seen the future as being to the right rather than to the left, no doubt some influence of constructing timelines in History class.

[11] See Lakoff and Johnson's *Metaphors We Live By* (14–21) for a host of embodied examples including happiness being up, being controlled being down, and virtue and depravity being up and down respectively.

I'm predominantly going to be focussing on metaphor as skills and experience transplanted to novel situations, but it's worth saying a little more on the importance of language for even this more physical metaphorical deployment before we continue as language and other cultural practices significantly impact upon the gestalts that we encounter. Our prior experience doesn't exist in some vacuum; as I've already established, our existence in communities of users has a profound effect on the ways in which we encounter our artefacts and this is no less true for the establishment of whole-composites. That a communal language conditions our experience of the world is at the heart of the critical and cultural theory of the mid-twentieth century to the present, from the influence of Ferdinand de Saussure's radical revisions of semiotics to Foucault's discussions of linguistic power and ideology; from Derrida's critique of binary oppositions and logocentrism to Judith Butler's description of the role of language in performative gender roles. Each of these thinkers, and the reams of theory and philosophy that they have influenced, are simply the most recent examples of a long history of thought about the power of language to condition our reception of the world. The postphenomenological philosopher Don Ihde draws on this history in his phenomenology of multistable visual effects,[12] alongside the "hermeneutic phenomenology" of Heidegger and Paul Ricouer,[13] in order to describe the ways in which raw sense perception is primed by the stories drilled into the observer by her prior experience:

> In a hermeneutic strategy, stories...are used to create an immediate...context; they derive their power of suggestion from familiarity or from elements of ordinary experience. The story creates a condition that immediately sediments the perceptual possibility. In untheoretical contexts, this has long been used to let someone see

---

[12] Ihde describes the phenomenological experience of viewing shapes and objects, such as the Necker cube, from the same vantage, but with a different perception. (The Necker cube is a 2D line drawing of a 3D cube that either appears to have its front face at the level of the paper with depth pushing "back" or its rear face at the level of the paper with its depth projecting "out." With a little practice you can switch between these different "views" of the same drawing, a multistable effect.) See chapter 4 of Ihde's *Experimental Phenomenology* for an extensive discussion of various effects of this kind.

[13] Hermeneutic phenomenology aims to interpret (rather than simply describe) the world; it recognises the effects of culture on perception and particularly on meaning which is never direct and must always detour through cultural influence.

something. Storytellers, mythmakers, novelists, artists, and poets have all used similar means to let something be seen. Plato, at the rise of classical philosophy, often paired a myth or fable with argument or dialectic. Within the context of the story, experience takes shape. (*Embodied* 61)

Elsewhere Ihde states the case more plainly: "Perception takes shape within and from the power of suggestion of a language-game. It sees according to language. This strategy is the basis of what has become known as hermeneutic phenomenology" (*Experimental* 61). Now, I don't want to make this point too strongly, to suggest that language plays the dominant or even one of the most significant roles in how we encounter familiar and novel objects; this is the over-dramatisation of the linguistic turn in cultural theory. But language does represent another way in which how we see the world can be affected by prior experience. In this regard, language can act as a repository for our knowledge about the world and a primer and shaper of future action in both explicit and subtle ways.

Stories matter both for our imagination and for our perception of things. The repetition of experiences can build up a "grammar" which we can apply to interpret future situations. Visual grammar is a simple example. For example, we've been trained by film, photography, painting, and particularly comics and other sequential art to draw connections between consecutive images, to tell a story. I want you to imagine three detailed paintings; take a couple of seconds to visualise each one in turn. First: a baby's head. Second: a chicken's egg. Third: a hammer. Picture this triptych and you should start to feel uncomfortable, because you start to turn the images into a narrative. There's no inherent reason to do so, but we have been trained to make metaphorical and temporal links between images – there's no story, and yet we somehow fear for this Platonic child out there somewhere, a colonisation of our thought through training affecting our perceptions.

Similarly, if I say: "imagine Antarctica" then most readers will now have a very clear image in their heads: snow and ice, a white desert, maybe some penguins or hardy seals, maybe a shape on a map or globe. And I'm guessing that you probably haven't been there, and yet that image, for all that it's just a composite of the hundreds and thousands of images that you've seen, it has a real effect on your understanding of the earth. Visual media, and the stories that we are encouraged to construct, invade our sense of the world. Susan Sontag, in *On Photography*, noted that we can no longer look at a sunset without what's really in front of us being judged in relation to the sunsets of the films, photographs, and paintings that occupy our minds as much or more than our real prior experiences of the phenomenon – "[c]ertain glories of nature...have been all but abandoned to the indefatigable attentions of amateur camera buffs. The image-surfeited are likely to find sunsets corny; they now look, alas, too much like photographs" (85). Sontag was concerned with how reality becomes distorted by the ways in which we saturate ourselves with images, but that's also a significant feature of how perception functions: the marshalling of past experience to produce a productive distortion and reduction of the immense totality of sense impressions that we access every instant of our lives.

Several recent experimental programs have also rekindled interest in the explicit role that language plays in perception, an area of research critiqued to the point of near extinction in the last years of the twentieth century. Landau et al.'s "The Influence of Language on Perception: Listening to Sentences about Faces Affects the Perception of Faces," for instance, demonstrates what it's title suggests, as does Gary Lupyan and Emily Ward's "Language Can Boost Otherwise Unseen Objects into Visual Awareness."[14] But my acceptance that language can structure perception and the emergence of gestalts shouldn't be mistaken for a full commitment to (what has become known as) the Sapir-Whorf hypothesis, strong linguistic relativism, or wholly social constructivist or constructionist models. Rather, as we'll go on to explore, I intend a weaker notion that goes beyond linguistic determinism to suggest that we take on a wide variety of strategies, often metaphorical, never simply linguistic, that structure the ways in which we respond to the world; language and stories play an important part, but no one strategy consistently dominates across encounters. That the stories culture tells us have an effect is an intuitive claim however: I'm primed to find dogs companionable, words in books important, and enduring monogamous love achievable and desirable, at least in part, by my cultural background. All of these things could no doubt manifest independently of my cultural priming, but equally undoubtedly my beliefs, perceptions, and investments of effort are also deeply affected by my linguistic and cultural conditioning.

Such concerns with language, media, and story-telling are familiar to the Humanities, if not always in quite these forms. What I want to now add to the roles of linguistic and more broadly cultural priming are the less explored embodied effects generated by both user and artefact in

---

[14] A nice discussion of the history of linguistic relativism, its relationship with Cognitive Science, and its slow return to plausibility can be found in Chris Swoyer's "How Does Language Affect Thought?"

order to demonstrate that priming is far from deterministic, far from simple, far from predictable, and absolutely present in our every experience of the world. These are another kind of metaphor.

3.2.1 Physical metaphors

As beings that evolved not just an innate intelligence, but also a staggering plasticity as a survival strategy to handle the pressures of a rapidly changing world, we have learned to turn to metaphor to overcome the limitations that stem from our lack of specificity. As stated earlier, humans were never the strongest or the fastest, but we became the smartest, where "smartness" isn't a measure of fact retention or IQ, but of the swiftness of our cognitive adaptation to novel situations followed by a continued refinement of successful techniques. The technologising of artefacts, therefore, is simply one example of our adaptive smartness. But our adaptability means that we are never experts in all areas, specialising only as the environment requires, and only developing those skills that are repeatably useful.

Our bodies, as Damasio noted, are the relative constants which broadly endure over time, not (on evolutionary timescales) our cultures or their conceptions, so it is unsurprising that we have evolved to conceive of the world as it relates to our bodily experience,[15] not the voguish abstractions of our era. As the evolutionary psychologist Henry Plotkin puts it: "[w]e may like to think that we can think of anything and in any way we choose. And perhaps we can, but doing so takes a great deal of hard work. For most people logic and mathematics are not 'easy'. Unconstrained, general, context-independent and domain-unspecific thought does not come naturally" (198).[16] Unconstrained thought, ideas without contexts, ideas that we cannot categorise are all troubling to us. This goes some way to explaining our dependence on metaphor, both in our abstract thought, as Lakoff and Johnson argue, but also in our

---

[15] This is not to align my thinking wholly with the "nativism" of, for instance, Jerry Fodor (*The Language of Thought*) or Steven Pinker (*The Language Instinct*) who think of our minds as remaining evolved for a different era (claims which seem wholly refuted by the extent of our neural plasticity). But I do want to recognise that *despite* our deep plasticity there remain enduring evolutionary effects, particularly around our experience of our bodies and the experience that they provide us of the world – evolution has shaped the kinds, limits, and capacities of plasticity; it came from somewhere.

[16] Plotkin is another nativist, but this is the first of two claims where I side with him; the second, his approach to the implications of evolution for epistemology, will be crucial to the final chapter.

attitudes to any unfamiliar thing. Metaphor, intelligence about one area deployed in another arena, provides practical constraints and enables us to deploy domains of expertise that we have evolved through repeated use, either culturally or genetically, in order to swiftly interact with, and swiftly adapt to, new aspects of the world.

The example of the move from page to screen is again useful here: for the reader with a lifetime of codex reading, moving to reading on an e-reader can be a dramatic (even, as we've seen, traumatic) shift. A significant component of this troubling is the requirement, and failure, of the deployment of metaphor; in the new digital equipment for reading the user interface and hypertextual database network can sit uncomfortably alongside the codex tradition. Hayles, for instance, describes the seemingly spectral materiality of digital texts:

> In the computer, the signifier exists not as a durably inscribed flat mark but as a screenic image produced by layers of code precisely correlated through correspondence rules, from the electronic polarities that correlate with the bit stream to the bits that correlate with binary numbers, to the numbers that correlate with higher-level statements, such as commands, and so on. Even when electronic hypertexts simulate the appearance of durably inscribed marks, they are transitory images that need to be constantly refreshed...to give the illusion of stable endurance through time. ("Print Is Flat" 74)

Elsewhere this leads Hayles to argue that "electronic text is a process rather than an artefact one can hold in one's hand" ("Deeper"). When a book becomes a temporarily instantiated process resting on multiple layers of code, rather than something neat to be held, there are clear implications for how we conceive of it. When we are first faced with an electronic text we are required to direct our visceral experience at some novel gestalt for us to begin an interaction, that is, we must make the artefact conform to being at least some sort of device to us if we are to use it. In this production we must initially draw on prior reading contexts and prior whole-composites from what we believe are similar artefacts, that is, we have a *heuristic* for reading that we hope will transfer. This strategy has mixed results.

If we encounter an object that needs to be hit, and we have a heavy object to hand, then we very simply draw on our history of experience with hammering. Text is only different in its complexity and the richness of the heuristic required for the engagement. The process of drawing meaning out of the intricacy of a page, however it is instantiated, requires a fairly stable set of restrictions on the kinds of actions that readers can attempt, and the aspects that they attribute meaning to. With poetry, for instance, we learn over repeated engagements that a line break can have meaning, but we ensure that this is not a part of our engagement with a novel; instead the line breaks are necessarily ignored. The rules for reading, of what should and shouldn't mean, are complex and must be flexible enough to respond to variations in content (e.g., in a novel where line and page breaks *are* put to the task of meaning). Part of the struggle with digitised work, then, stems from our attempts to establish new embodied and hermeneutic strategies for apparently incorporeal reading spaces by drawing on the learned interactions with more simply manifested print.

The digital document drives us to our history of print, and to writing more generally (both of which do largely hold as good model metaphors for the engagement), but its appearance on an electronic artefact also sends us to our experience of various prior screens:

> Readers come to digital work with expectations formed by print, including extensive and deep tacit knowledge of letter forms, print conventions, and print literary modes...At the same time, because electronic literature is normally created and performed within a context of networked and programmable media, it is also informed by the powerhouses of contemporary culture, particularly computer games, films, animations, digital arts, graphic design, and electronic visual culture. (Hayles, *Electronic* 4)

The seeming range of potential (and desirable) encounters is at first massively expanded to include this diversity of prior contexts, experiences that we might draw upon in order to attempt action – all of these things are suddenly part of both what "book" and "digital/electronic text" can mean. Hayles here identifies the challenge in producing a new gestalt for the artefact, a shape that doesn't settle easily and that can only be refined through interaction. Those first uses, if they are to have any chance of success, must heavily rely on our prior activities – I can't just refer to the codex which has taught me how to read, I need to think of other electronic devices to give me a chance to understand how to use buttons, how to navigate menus, how carefully I need to treat the artefact, etc. The whole-composite that starts to form is initially a mess, bringing in a host of co-present features that aren't really there, setting me up to be thwarted, but it also gives me hope of success.

Lakoff and Johnson provide an example that might help elucidate this further, a description of how a change in conversation can be understood by deploying a concept from another realm:[17]

> As we experience a conversation, we are automatically and unconsciously classifying our experience in terms of the natural dimensions of the CONVERSATION gestalt: Who's participating? Whose turn is it?...What stage are we at? And so on. It is in terms of imposing the CONVERSATION gestalt on what is happening that we experience the talking and listening that we engage in as a particular kind of experience, namely, a conversation. When we perceive dimensions of our experience [of an exchange] as fitting the WAR gestalt in addition, we become aware that we are participating in another kind of experience, namely, an argument. It is by this means that we classify particular experiences, and we need to classify our experiences in order to comprehend, so that we will know what to do. (*Metaphors* 82–83, capitals in original)

The CONVERSATION gestalt, a typical collection of elements that we encounter and name, is born of experience with conversing and we are able to form a particular understanding of its distinguishing parameters so that when we meet an experience which seems to fit those elements then we can act with a predetermined heuristic or set of gestures which enable us to successfully negotiate the particular instance. Our ability to remember prior interactions with whole-composites allows us to simplify and quickly adapt to new situations that cause us to perceive the same or similar gestalts. The gestalt of WAR can be overlaid onto the CONVERSATION gestalt when what we encounter changes, and this enables us to, if necessary, become combative, give up or lose ground, attack weaknesses, and shoot down suggestions, etc. (4). The gestalt that we have for a conversation alone is not enough to get us through the particulars of the new instance, so knowledge from another realm is included to bolster the range of options and responses available to us. Electronic reading (as with any novel engagement with a new artefact) operates similarly: we have a default heuristic for book reading (a set of gestures for engagement similar to Lakoff and Johnson's interest

---

[17] For Lakoff and Johnson, the collection of elements into concepts that are experienced as unitary are the gestalts that they're describing. For my purposes, "gestalt" refers to any whole-composite, but I'm focused in this chapter on artefacts rather than Lakoff and Johnson's linguistic concerns.

in conversation) and a whole-composite that we expect to encounter when reading a printed book (a gestalt artefact). These collections of aspects, experienced as unities, enable us to identify an engagement by referring to our prior experiences. Encountering an electronic text, we are then initially forced to apply these same paradigms, but electronic reading is capable of, and promotes, interactions such as clicking, scrolling, zooming, and swift travel between content, etc., which don't fit in with our printed book experience. We must then suddenly flail to find a suitable model from elsewhere to get us through the experience, one that can be grafted onto the gestalt that we first applied. Sometimes the search is brief, and we settle on and begin to redefine a prior relation almost immediately; sometimes we must swiftly cycle through options as the new engagement is dramatically dissimilar to our roster of previous experiences; and sometimes an interaction is even more subtle, fooling us into thinking that one or two of our past gestalts and heuristics are more than enough to conquer the new instance and then surprising us, maybe weeks later, with their unsatisfactory ability to aid comprehension of some required aspect of the thing.[18] This period of uncertain and unrelentingly devicive use, whilst new functional gestalts, and their attendant gestures, are forming is clearly not conducive to cognitive efficacy, and we might suspect that anyone for whom this time was excessive, too often witnessed in others, or never conquered in themselves, may not be disposed to championing the new equipment.

3.3 The problem with over-relying on metaphor

Lakoff and Johnson's discussion of structured gestalts being altered by metaphors drawn from other realms of experience also offers us both

---

[18] The classicist James O'Donell offers a folk phenomenological description of the development of a new heuristic for thinking online:
it's my fingers I notice... [W]hen you've asked a really interesting question... it's a physical reaction, a gut feeling that I need to start manipulating (the Latin root for "hand," manus, is in that word) the information... to find the data that will support a good answer... The sign of thinking is that I reach for the mouse and start "shaking it loose" – the circular pattern on the mouse pad that lets me see where the mouse arrow is, to make sure the right browser is open, get a search window handy. My eyes and hands have already learned to work together in new ways with my brain – in a process of clicking, typing a couple of words, clicking, scanning, clicking again – which really is a new way of thinking for me (192).
New ways of thinking go hand-in-hand with new ways of acting, and, importantly, vice versa – it is worth developing the new.

warning and explanation as to why it is not enough simply to rely on a combination of our codex and computing gestalts and hope for the best:

> Having a basis for expectation and action is important for survival. But it is one thing to impose a single objectivist model in some restricted situations and to function in terms of that model – perhaps successfully; it is another to conclude that the model is an accurate reflection of reality... *To operate only in terms of a consistent set of metaphors is to hide many aspects of reality*. Successful functioning in our daily lives seems to require a constant shifting of metaphors. (*Metaphors* 221, emphasis in original)

Lakoff and Johnson's work can again be extended to our discussion of technological artefacts. If we keep on metaphorically deploying sets of principles from other areas of our experience in order to understand a new interaction, then we will unavoidably hide many useful aspects of the object in our production of some new whole-composite. If we want to really understand and effectively use an e-reader or tablet, then we cannot stay thinking of it as a kind of codex crossed with a computer. This will prevent it from becoming a technology and reliably remaining as such; there will always be moments of profound rupture as we're disappointed by what it cannot do instead of appreciating what it might be ideally suited for (and note that this is true of any interaction with a new artefact). When the method of the new engagement closely overlaps with activities associated with our experience of another type of object, it can cause a cognitive dissonance amongst new users as similar activities, though impossible to be identical, can be attempted with both. If we get stranded wondering why an e-reader isn't behaving like a codex, then we might certainly feel that the new medium is deficient, that it may be detrimental to our thinking because we can't use it as we want to. This is a failure of metaphorical support, an over-reliance on metaphor, and one that can perversely evolve to be built into the fabric of the thing. As Lakoff and Johnson describe it, in terms of a form of linguistic determinism, "[m]etaphors may create realities for us, especially social realities. A metaphor may thus be a guide for future action. Such actions will, of course, fit the metaphor. This will, in turn, reinforce the power of the metaphor to make experience coherent. In this sense metaphors can be self-fulfilling prophecies" (*Metaphors* 156). In much the same way, the codex acts as a vital structuring metaphor for the emergent experiences of the e-reader and the digital text. But an over-reliance on this metaphorical filter cannot help but lead to our treating the new form as if it should act in terms of the old. This is problematic for several reasons, the most significant of which are that it restricts the new form from becoming all that it might be, and it impedes our understanding of the thing and the beneficial uses that it might provoke.

We'll pick up on this more in Chapter 4, but it's worth saying here that insufficient metaphors will in this way deny us access to expertise and technological use. True expertise must be about experiencing things as closely as possible as what-they-are, not in terms of some other thing; the reliance on metaphor will keep us as amateurs unable to see the true affordances of the artefact that we are working with. When we are first being taught how to use something our teacher might say "it's a bit like..." or "just pretend that you're..." But when you're refining a skill at a higher level the physical (if not linguistic) metaphors drop away: "I've got to keep playing through the ball for accuracy and power"; "I need to stop rotating my wrists so I can drive the nail more cleanly"; "I need to track to where the target will be before I fire over this distance." Each of these instances is sensitive to the activities and the artefacts *as they are*, not in terms of some other experience.

We have already seen some of the failure of an over-reliance on metaphor in the reports that have been discussed from Chapter 1: much of the resistance to the new reading equipment can be attributed to a misunderstanding of what they are in comparison to the body of an idealised codex. Ben Vershbow and Dan Visel, writing at the *Future of the Book* blog well before the release of the first Kindle, specifically articulated this problem in relation to reading online, and their exchange is worth spending a little time with, beginning with Vershbow's initial post:

> A plant in a container grows differently than a plant in open soil. The roots conform to the shape of the pot. Similarly, our very notions of reading, of books, of knowledge classification are defined by the pot in which they grew. The texture of paper, the topography of the library, the entire university system – these were defined by restraints. Physical, economic, etc. And to a significant extent they are artifacts of their times...The computer, too, in its current stage of development, is an artifact of the paper book, the typewriter, and the supercomputer terminal. These define the "pot" in which the computer has grown. And so far, the questions about online "reading" are defined by this cramped root structure. Even though the pot has shattered, we continue to grow as though the walls were there. ("The Cramped Root: Worshipping the Artifact")

Here Vershbow identifies the enduring metaphors that continue to shape the gestalt artefacts and activities of contemporary home computing.

Later, Visel responded to this post discussing the "arrogance" of assuming that we should read, for instance, Wikipedia like we do a printed encyclopaedia, and in his discussion he invokes another established metaphor, the "horseless carriage" (which Vershbow also mentions), used as a way of understanding the newly invented automobile:

> [I]t's a human response to compare something new to something we already know, but often when we do this, we miss major formal differences... [W]hat's happened with the Wikipedia is that a new species of text has arisen and we're still wondering why it won't eat the apples we're proffering it. We judge it by what we're used to, and everyone loses. Were you to judge a car by a horse's attributes, you wouldn't expect to have an oil crisis in a century. ("Learning to Read")

These commentators repeat Lakoff and Johnson's concerns about an over-reliance on metaphor, but specifically in terms of e-reading: to continue to judge the new by the old means that we still lose what's good about the old thing (it may not be present anymore), but we also lose what might be excellent about the new. Vershbow's point about the constricting pot is vital – the codex evolved under the constraints and provocations of its environment directly resulting in its materiality (something that we'll pick up on in the final chapter). If we treat the new forms as simply electronic codices, then we inherit not only the useful structuring metaphors, but also a set of limitations in deployment and practice from another era. And when people experiment with new forms of production or consumption, and Wikipedia is a great example of both (particularly in comparison to the imposing physicality of off-screen encyclopaedias), they can often seem bereft in their growing pains rather than a productive move towards eschewing unnecessary constraints – in the case of Wikipedia constraints such as storage, linking, portability, authorial bias, and decisions regarding suitability of materials can and have been ameliorated to a greater or lesser degree by the new form. As Bolter puts it, the "shift from print to the computer does not mean the end of literacy. What will be lost is not literacy itself, but the literacy of print, for the electronic technology offers us a new kind of book and new ways to read and write" (2). Metaphor, ironically, tends to be innately at odds with the new.

3.3.1 Skeuomorphism

The "new book" that Bolter saw coming became established in the form of a text maintained on a handheld device rather than on any desktop or laptop. I suspect that, besides the convenience, the connotations of "book" are deeply tied to a lengthy text that is contained in a portable form. Though we can stretch such terms, though they might, even when stripped down this far, be immensely plastic, if they are pushed too far or too hard then they will no longer seem to fit even those idiosyncratic gestalts that we label "book" for ourselves. But at the same time, if we are too limited by our prior experiences, if we don't push at them at all, then we might still end up with a car that doesn't like apples. In fact, Lakoff and Johnson's concerns with metaphors as "self-fulfilling prophecies" again have a relevant physical analogue that represents another real impediment to the sustainable development of our interactions with new equipment.

In evolutionary terms, the forms that I want to discuss are called "vestiges," residual elements of evolved adaptations which were a part of the species' past suitability for an environment; "[t]he history of the species lives on in the modern species, and one of the most important pieces of evidence for this lies in vestiges of previous adaptations that may prove redundant, or useless, or even maladaptive, today" (Donald, *Origins* 121). The coccyx in humans, for instance, the extended "tail bone," is a vestige of the true tail that was a part of our distant simian ancestors. Our artefacts might also be thought of as containing vestiges, and just as with their biological counterparts they persist because the environment doesn't enforce pressure enough to select them out. With our artefacts, of course, *we* create the selective pressures, and we aren't always neutral in the forces that we develop. This also means that our cultural quirks, such as an over-reliance on metaphorically deployed gestalts and heuristics, can create pressures that *promote* vestiges in a way that is rarely found in the natural world.

The term for such vestiges within artefacts, a term drawn from archaeology, is "skeuomorph": "Simply put they are carryovers from an older technology or way of doing things that had value, and are retained as a semblance, and expectation. Characteristic of changes in technology, they confer a kind of luster. The technological reason for the feature has gone, but you expect it – it *completes* the object" (Taylor 153, my emphasis). Taylor's example is that of a bottle of wine, or, more specifically the dent in its base: "When wine bottles were blown, there was no alternative: the molten glass bubbled out like a long balloon with a rounded end; this base was then flipped inside out as the bottle was set down to cool, producing the level circumferential basal ring that would allow the bottle to stand upright" (153). But now wine bottles aren't made like that, they're formed of two halves within a mould before being joined; there's no need for the dent.[19] But the bottle would be somehow incomplete without it; its presence is somehow more "authentic," and so it stays.[20]

Skeuomorphs are readable vestiges, hangovers from the deployment of prior gestalts and heuristics in the formation of new interactions that are written into the things themselves. They can often be useful to new users, acting as readily apparent material metaphors and enabling the adoption of new equipment more smoothly (as with any other metaphorical usage) by suggesting modes of engagement associated with the gestures and milieu of more familiar tools. As Hayles notes, "skeuomorphs [can act] as threshold devices, smoothing the transition between one conceptual constellation and another" (*How We Became Posthuman* 17).

But skeuomorphs, as with an over-reliance on metaphors of any kind, can also preserve unneeded restriction. Take, for instance, the features of the software that Apple bundled with the iPad and iPhone until 2013's iOS 7 designed by Jony Ive. Adam Greenfield, a former user-interface designer for the mobile telephone company Nokia, blogged about the cognitive dissonance he felt in using the software on his new iPhone 4:

> The iPhone and iPad...are history's first full-fledged everyware (*sic*) devices – post-PC interface devices of enormous power and grace – and here somebody in Apple's UX shop has saddled them with the most awful and mawkish and flat-out *tacky* visual cues...Dig...the page-curl animation (beautifully rendered, but stick-in-the-craw wrong) in iBooks. Feast your eyes on the leatherette Executive Desk Blotter nonsense going on in Notes. Open up Calendar, with its twee spiral-bound conceit, and gaze into the face of Fear. What are these but misguided coddles, patronizing crutches, interactively horseless carriages?...You give up the tangible, phenomenological isness of the book, and in return you're afforded an extraordinary new range of capabilities. Shouldn't the interface, y'know, reflect this? ("What Apple Needs to Do Now")

As a designer himself, Greenfield sees these user-interface choices, with their modelling of physical corollaries, primarily as tacky conceits, but

---

[19] And, as Taylor notes, in France the feature is called *"le voleur* ('the thief') because without it there would be more wine" (153).

[20] In this regard, Hayles sees skeuomorphs, the histories of artefacts, and the networks that they call into being (e.g., cars bringing about road networks), as complex foldings of time into objects (*How We Think* 89).

more importantly he also feels that by giving up on the specific materiality of the equipment that the new software hopes to emulate, its "isness," we shouldn't also forget the new range of capabilities that this allows; the differences, the gains and losses, need to be revealed, not occluded in a desperate bid to diffuse them. If Greenfield (and others[21]) thought that we could look past such skeuomorphs, then he would be less inclined to portray them as problematic. But these concerns for the interface's reflection of the new practices available stems, in part, from a fear that some users, maybe a majority, will not move on, will only hope to use a new thing in an old way, and a way that is not only short-sighted, but that the equipment may also not be able to live up to, resulting in its being rendered as a failure rather than a source of opportunity.

iOS 7 marked a change in mainstream digital software. Until iOS 7, Apple were the most prominent exponents of skeuomorphic user-interface design. Ive's redesign, however, abandoned real-world cues, at least in part in recognition that he was now free to do so: "we understood that people had already become comfortable with touching glass, they didn't need physical buttons, they understood the benefits...So there was an incredible liberty in not having to reference the physical world so literally. We were trying to create an environment that was less specific" (Molina & Bravo, "The Man Behind Apple's Magic Curtain"). For Ive, skeuomorphs had served their purpose and brought enough people on board. Opinion at Amazon, however, seems more mixed. The first generation of Kindle was wedge-shaped so that it felt something like an open paper book in the hand,[22] but it's now become a flat tablet. And yet, if you want to save a page (on its skeuomorphically named "Paperwhite" range) then you still turn down (the representation) of a "page's" corner – the vestiges of print culture still persist.

Though the inhibition of opportunity written into artefacts can certainly be problematic for the adoption of new devices, more broadly the reduction of our sensory experience through metaphor is essential. We can only encounter so much, and with so many objects in our experience, and so many new artefacts to explore, reductive metaphor can provide

---

[21] For more on this discussion see Clive Thomson's discussion at *Wired* "Retro Design is Crippling Innovation."

[22] Jeff Bezos said at the time of the launch of the original Kindle that "it must project an aura of *bookishness*; it should be less of a whizzy gizmo than an austere vessel of culture. Therefore the Kindle...has the dimensions of a paperback, with a tapering of its width that emulates the bulge towards the book's binding" (Levy, "The Future..." 57).

a compelling way of navigating first encounters and establishing useful whole-composite objects. E-reader design, and phone and tablet software, remains in an unstable balance between drawing on and fighting against centuries of use of the codex, against choices long made, and in this competition we see skeuomorphic tics manifest and disperse as designers attempt to both woo and shape the user with offerings of familiarity and the promise of new potential. All of this functions as evidence for metaphor's capacity for structuring the gestalts that we encounter.

3.4 Embodied suggestions

So far in this chapter, we've established that artefacts aren't encountered as a combination of individual parts, but instead as a whole-composite or gestalt, a mutable complex of elements that are always experienced as a perceptual unity. We've also explored how the nature of that unity is actuated by our prior experiences and our cultural conditioning, by various metaphorical redeployments of knowledge which enable us to encounter novel objects and rapidly turn them to use, and some of the implications of this for the development or inhibition of expertise. I want to end the discussion of encountering artefacts with two interrelated factors that further impact upon gestalt constructions before going on, in Chapter 4, to consider the ontological base behind the encounter of whole-composites. Both of these factors are tied to the "suggestions" provoked by the meeting of the object and the user's embodiment: affordances and our experience as beings who spread cognition (and its impacts on perception) over brain, body, and environment, something captured by recent threads in Cognitive Science and Don Ihde's postphenomenology. While embodiment is significant for our understanding of the power of metaphor, it is typically the most important shaper of the composite objects that we encounter through its involvement in cognition and perception.

3.4.1 Artefact embodiment: affordances

In *The Ecological Approach to Visual Perception*, the psychologist James Gibson developed his concept of "affordances." Gibson saw affordances as the action-facilitating aspects of something in an environment, what a thing "offers" an organism "what it provides or furnishes, either for good or ill... [An affordance] implies the complementarity of the animal and the environment" (*Ecological* 127). The nature (and existence) of affordances has been contested,[23] but I side with Anthony Chemero's

---

[23] See Chemero 105–161 for a full discussion of the debate around Gibson's work on affordances.

interpretation in *Radical Embodied Cognitive Science* that an affordance is real only at the meeting of animal and world (139–147), that is, that whilst an affordance is a real feature of an object (it continues to exist whether or not it is perceived by anything) that feature can also only be understood with regard to how it is encountered.[24] This seems to accord both with experiment[25] and with Gibson's original idea. Gibson offers the example of a surface for walking on, saying that if a surface is horizontal, flat, extended in comparison to the size of the organism, and rigid enough to support its weight "then the surface affords support... Note that the four properties listed – horizontal, flat, extended, rigid – would not be physical properties of a surface... As an affordance of support for a species of animal... they have to be measured relative to the animal" (127). Affordances, therefore, aren't the same as the perceptual aspect of a thing (which may or may not be accurate), but are instead, rather, real properties of the object that exist only in relation to their being actionable in use.[26] A codex, for example, affords reading whether or not someone is there to read it, but readability isn't a physical property of an object independent of a reader. Gestalts and heuristics are therefore partially formed from affordances, and also swiftly changed in response to their reality – if I perceive that a chair is made from connected slats of wood and affords support, then I am substantially surprised when in fact the slats aren't glued together; my gestalt for that chair (and for chairs more broadly), and my heuristic for sitting, are altered as I become wary.

Donald Norman, in his guide for designers *The Design of Everyday Things*, uses Gibson's notion of affordances in order to consider the readable properties of materials and artefacts,

> primarily those fundamental properties that determine just how the thing could possibly be used... Glass is for seeing through, and

---

[24] In the language of the next chapter we will be able to say that affordances are real qualities of sensual objects, but we have a little way to go before this can be explained fully.

[25] Again, see Chemero 105–161 for a discussion of the experiments that have explicitly attempted to define or explore the nature of affordances, including those undertaken by Chemero himself, see for example Charles Heyser and Anthony Chemero, "Novel Object Exploration in Mice: Not All Objects Are Created Equal."

[26] "The observer may or may not perceive or attend to the affordance, according to his needs, but the affordance, being invariant, is always there to be perceived. An affordance is not bestowed upon an object by a need of an observer and his act of perceiving it. The object offers what it does because it is what it is" (Gibson, *Ecological* 139).

for breaking. Wood is normally used for solidity, opacity, support, or carving. Flat, porous, smooth surfaces are for writing on... When affordances are taken advantage of, the user knows what to do just by looking: no picture, label, or instruction is required. (9)

Such awareness of material affordances, useful because born of experience, are just the kinds of guide that skeuomorphic user-interface design draws upon – we know how the real world is supposed to work and at least partly independently of our cultural training. Paper, in the terms that Norman offers here, affords both inscription and reading; there is a cultural component to both activities, of course, but there is also an important material reality that affords, and indeed has meaning, on its own terms. As the theorist and book artist Joanna Drucker puts it,

> [t]he force of stone, of ink, of papyrus, and of print all function within the signifying activity – not only because of their encoding within a cultural system of values whereby a stone inscription is accorded a higher stature than a typewritten memo, but because these values themselves come into being on account of the physical, material properties of these different media. Durability, scale, reflectiveness, richness, and density of saturation and color, tactile and visual pleasure – all of these factor in – not as transcendent and historically independent universals, but as aspects whose historical and cultural specificity cannot be divorced from their substantial properties. No amount of ideological or cultural valuation can transform the propensity of papyrus to deteriorate into gold's capacity to endure. (*The Visible Word* 45–46)

Culture can, and does, program our reception of materials, artefacts, and objects, and we often act on cultural impulses until we learn for ourselves the true affordances of a particular substance or object. But what Drucker really brings out here is that culture is never enough – materiality always means to us as embodied beings, it always speaks, and in large part through its affordances, what it facilitates for us. It is those affordances which we need or which we cannot ignore that stand out most and, again, cause us to reduce our experience of a whole to its most relevant aspects (whilst still experiencing it as united).

Similarly to the nostalgic draw of skeuomorphs, beneficial affordances of one object can blind us to the potentials in another. Derrida, for instance, during a discussion of his own play with the writing space, argued that the ubiquitous presence of the computer, and absence of paper, both in the world and in his hands, would have altered the work that he produced: "I think that the typographical experiments..., particularly the ones in *Glas*, wouldn't have been interesting to me any more; on a computer, and without those constraints of paper – its hardness, its limits, its resistance – I wouldn't have desired them" (*Paper* 47). There is something in the paper itself, Derrida is saying, that prompted in him, that afforded certain kinds of thinking, and therefore certain kinds of work. There is also, absolutely, a wave of cultural assumptions that surround paper, particularly the paper in printed books, which speaks to, and to some degree creates, "its hardness, its limits, its resistance." But to a greater or lesser extent those cultural biases must be able to cohere relative to the objects that they cluster around, as exemplified in the quotation from Drucker above. Derrida realised that the computer silences paper and its potentials, at least as much as paper's power to silence the potential of the digital.

Affordances, then, are the aspects of the world that we cannot escape at the moment of use. We might initially misperceive what an object can do because our conditioned gestalt is insufficient, but there is always a revelation at the point of action. And when an object's true affordances, or at least the ones that we experience (we always miss more than we encounter), when they come to light they become a dramatic shaper of our future gestalts and heuristics for seemingly similar items.

3.4.2 User embodiment: spreading cognition

The simplest way in which materiality affects how we encounter an artefact, however, is through our own embodiment. I cannot see a mobile phone's constituent atoms; I cannot see the electricity grid or Wi-Fi microwaves that it may be connected to; I cannot see inside it when it's closed, or behind it when I face its front. These raw facts of phenomenological experience are the true fundamental behind the production of the object as a whole-composite – we predominantly live in the chronological meso-scale of embodied human experience, barred from the micro, the macro, the past, and the future by our gross forms and the limitations of our perception.

We've also already heard from Husserl about our meso-lives, about how our embodied experience enables us to perceive an object as an object despite it's only ever appearing from one aspect, one adumbration. The anticipated and co-present other faces of a thing are part of what makes something what-it-is and what-it-is-for-us (perception is never the thing itself even as it must be partly parasitical on some real thing). Husserl had some faith that we can find the object itself from the way it appears in the world, but I will continue to argue that we just encounter a never-finalised and externally-influenced gestalt with the thing itself in a constant process of escaping from our perception, as evidenced by the continued possibility of surprise in even the most experienced and expert of interactions.

To conclude this discussion of encountering artefacts as whole-composites I want to (very) briefly outline some significant research areas in contemporary Cognitive Science, their insights into the influence of the body on the formation of perception and cognition, and how they might be of use to our understanding of technological artefacts. These areas are Distributed, Embodied, Extended, Embedded, Enactive, and Situated Cognition, Radical Embodied Cognition, and Postphenomenology (I will spend a little more time on Postphenomenology as it is the area most explicitly relevant to discussions of technology and the majority of this work accords with, extends, or challenges aspects of the postphenomenological position). As will become clear, each of these research programs has affected the conception of technology and the impact of embodied use on perception and cognition outlined in this book. It is hard to discuss any of these approaches in isolation as insights from each area continually condition the rest; what follows is therefore a somewhat artificial separation, but it allows me to offer some specific cases of the effects of embodiment on gestalt perception. Chemero's work, as we'll see, goes some way to uniting the most significant features of these fields, but his agenda in *Radical Embodied Cognitive Science* is different from my own here (aiming to establish a practical Cognitive Science research agenda); after this section I will instead cumulatively refer to these united concerns as "4EDS Cognition" (embodied, extended, etc.) to highlight the ways in which cognition is spread, building on the already and increasingly used term "4E Cognition."[27] "4EDS" will act as both promissory note that the cognition I am discussing isn't simply a process that goes on in the head and a reminder of the impact of body and world on the whole-composites that we can therefore form.[28]

---

[27] See, for example, Richard Menary "Introduction to the Special Issue on 4E Cognition."

[28] Malafouris offers an excellent extended introductory bibliography of writers who could be considered under the 4EDS umbrella; see *How Things Shape the Mind* (57).

3.4.2.1 Distributed cognition

Distributed Cognition sees some, maybe most, cognitive processes as spread over a system of cognisers and objects in the environment.[29] For instance, Edwin Hutchins' work of cognitive anthropology *Cognition in the Wild*, a foundational text for Distributed Cognition, offers an extended study of navigation on board large naval ships and the roles undertaken by various navigation personnel and the equipment that they deploy in their activities. Hutchins sees each of these actants, human and nonhuman, as contributing to a distributed computational system that can be seen as cognising in its own right.

Distributed Cognition is relevant to our discussion here because a significant aspect of the field sees technological artefacts as playing a real role in cognition – there are cognitive tasks which can be shared across other people and other things, and that sharing is another approach to both the importance of communities in structuring experience and the role of Incorporation. We can also take from this that Distributed Cognition is compatible with different users encountering different features of the same experience, reducing situations or objects to particular aspects and with their current roles dictating the most salient aspects to attend to – the distribution of our cognition affects the features that we cohere into the gestalts that we encounter.

3.4.2.2 Embodied Cognition

Embodied Cognition also holds that cognition doesn't just occur in the brain, emphasising the vital role that the body plays in conditioning our experience – we think what we think, and we *can* think what we think, at least in part due to our physicality. We've already considered several examples of the importance of the body in cognition (e.g., Goldin-Meadow's work on gesture or Davoli et al.'s work on touch in reading in Chapter 1), and for the formation of gestalts (above), so I'll offer a final example more concerned with practice.

Recent cognitive and neuropsychological work investigating the specificities of how we use, position, and perceive our hands in action suggests that they can have implications for our attitude towards our

---

[29] It is worth noting that Distributed Cognition is sometimes seen as an umbrella term for the rest of the approaches discussed here. See for example the recently funded *A History of Distributed Cognition* project led by Douglas Cairns; the project team are variously interested in each of the fields here and have opted for the term "Distributed" to cover their concerns.

environment, that is, our hands in use can affect our attention in the world. Ed Symes et al.'s "Grasp Preparation Improves Change-Detection for Congruent Objects," for instance, describes how the research group presented their subjects with a cycle of two images, one then the other repeated several times, both portraying two dozen fruits and vegetables of various sizes, but with one similarly sized object changing (i.e., all objects stayed the same bar an orange changing for a similarly sized apple, hard to spot in the first few exchanges). The subjects were asked to indicate when they had identified which object altered between the images by either (a) squeezing a handle in their fist (power grip) or (b) pinching a small switch (precision grip). When the object that altered matched the grip that had been "primed," then the latency of the response decreased significantly, that is, when an apple changed to an orange those participants which had to register the change with a power grip (the grip that would be deployed when interacting with the changing objects in real experience) outperformed those subjects who had to indicate the change with a precision grip (better suited to a grape or blueberry). Priming the grip measurably affected the subjects' perception of a scene; they were more ready to see things that they were primed to interact with.

Such research offers another take on "story-telling," Ihde's "hermeneutic strategies," affecting what we perceive; the actions of our bodies give cues for further action, they tell their own story. More broadly, Embodied Cognition, as a research agenda, aims to learn and empirically test the variety of ways in which embodiment affects cognition and perception,[30] and these concerns should make it central to any contemporary philosophy of technology. With regard to gestalts, an Embodied approach continues our phenomenological concerns with how the realities of our material existence impact upon the objects that we encounter.

3.4.2.3 Extended Cognition

Extended Cognition again moves the site of the cognitive process away from the brain, arguing that cognition isn't restricted to the boundaries of skin and skull but spreads out into the world.[31] Where Distributed

---

[30] For more on Embodied Cognition, see, for example, Varela et al. *The Embodied Mind*; Shaun Gallagher *How the Body Shapes the Mind*; Lakoff and Johnson *Philosophy in the Flesh*.

[31] For more on Extended Cognition see for example Clark *Supersizing the Mind*; Clark and Chalmers "The Extended Mind"; Richard Menary *The Extended Mind*.

Cognition, at least in work like Hutchins' *Cognition in the Wild*, tends to explore large systems of actants and the cognitive states of such systems, Extended Cognition tends to focus in on the role of objects in the world for individual human cognisers. We saw work from Extended Cognition in Chapter 2 from Clark and from Dotov et al., indeed the criterion of Incorporation as a mark of the technological is reliant on external implements being taken "on board" and being thought with and through. We've therefore covered a number of academic examples, and we'll continue to discuss fields concerned with cognition's breaking of the brain and skin boundary in the next sections, so here I'll simply offer a folk report of the phenomenon from the former-chef, now food writer and broadcaster, Anthony Bourdain:

> Mise-en-place is the religion of all good line cooks. Do not fuck with a line cook's "meez" – meaning their set-up, their carefully arranged supplies of sea salt, rough-cracked pepper, softened butter, cooking oil, wine, back-ups and so on. As a cook, your station, and its condition, its state of readiness, is an extension of your nervous system – and it is profoundly upsetting if another cook or, God forbid a *waiter* – disturbs your precisely and carefully laid-out system... I worked with a chef who used to step behind the line to a dirty cook's station in the middle of the rush to explain why the offending cook was falling behind. He'd press his palm down on the cutting board, which was littered with peppercorns, spattered sauce, bits of parsley, breadcrumbs and the usual flotsam and jetsam that accumulates quickly on a station if not constantly wiped away with a moist side-towel. "You see this?" he'd inquire..., "That's what the inside of your head looks like now. *Work clean*!" (58–59, emphasis in original)

Extended Cognition researchers, like the head chef of Bourdain's story, recognise that thinking is taken out into the world; the "inside of your head" includes the place that you're working and the things that you work with. Again, Extended Cognition seems to me to be of central interest to the contemporary philosophy of technology, giving us a source of empirical study and phenomenological report for the ways in which our artefacts can take an intimate role in our experience of the world, as true constituent components of cognition.

3.4.2.4 *Embedded/Enactive/Situated cognition*

Embedded, Enactive, and Situated approaches to cognition are closely related, and often used interchangeably (particularly Enactive and

Situated). Such approaches see cognition and perception emerging out of an interplay between the subject and their current environment, with Enactive and Situated Cognition tending to focus on the subject's action within a situation and Embedded Cognition more typically focusing on passive provocations from cultural and environmental concerns. The key insight is that the environment is not just the "stage" that cognition plays out in, but is, instead, rightly considered as an active element in the cognitive process.

With regard to the formation of whole-composites, our embeddedness in an environment and situation both restricts and enables the kinds of actions, conceptions, and perceptions available to us (as seen in Gibsonian affordances, e.g., or in a situation's dictating the adumbrations of a thing that I perceive). Embedded Cognition in particular has therefore been seen as closely allied with Embodied Cognition, and the two are often seen discussed together as EEC, that is, how our physical instantiations unite with the material environment that we find ourselves in in order to coproduce perception and cognition.[32] What is significant for us here is that every technological interaction takes place in a situation that impacts upon the experience, continually offering feedback, restrictions, and afforded actions.

3.4.3 Radical Embodied Cognition

In *Radical Embodied Cognitive Science*, Anthony Chemero considers the influence of the above fields on contemporary Cognitive Science[33] and aims to further, to radicalise their arguments. The radical approach amounts to two distinct claims:[34] (i) cognition is non-representational

---

[32] For more on EEC and Embedded, Enactive, and Situated Cognition see David Ward and Mog Stapleton "Es Are Good: Cognition as Enacted, Embodied, Embedded, Affective and Extended"; Hillel Chiel and Randall Beer "The Brain Has a Body: Adaptive Behavior Emerges from Interactions of Nervous System, Body and Environment"; Clark *Being There: Putting Brain, Body and World Together Again* (particularly 97–102); and Varela et al. *The Embodied Mind*. An interesting discussion of Embedded Cognition also appears in Mark Rowlands' chapter on 4E cognition in *The New Science of Mind* (51–84).

[33] See Chemero 17–42 for a thorough history of the positions and distinctions of embodied approaches in Cognitive Science.

[34] Chemero phrases it as three for the argument that he wants to make, but the first and third can be conflated – both refute the validity of cognition as representational.

and non-computational[35] and (ii) approaches to cognition which explore the importance of embodiment require a particular set of explanatory tools, including recognising human interactions in their environment as comprising dynamical systems (29). The first position is somewhat beyond our discussion here, but it rests on Chemero's argument that explaining cognition "in terms of its representations invites the anti-extended [cognition] claim that it is the represented environment, and not the environment itself, that is part of the cognitive system" (31). Chemero believes that if humans rely on mental models of their environment, rather than directly entering into a cognitive system *with* that environment, then the research described above cannot fully endorse its own claims – in this view radicalised 4EDS approaches to cognition must rest upon cognition occurring in direct contact with the world (if there is no directly produced system then there must be mental modelling).

The second claim is related and also at the heart of the argument of this book. It can also give us an important link between the role of embodiment in Cognitive Science and Postphenomenology:

> In radical embodied cognitive science... [a]gents and environments are modelled as non-linearly coupled dynamical systems. Because the agent and environment are non-linearly coupled, they form a unified, nondecomposable system, which is to say that they form a system whose behaviour cannot be modelled, even approximately, as a set of separate parts. (Chemero 31)

For Chemero, embodied humans constantly act in concert with their world and cognise in direct contact with it (i.e., cognition occurs in the dynamic systems that they form) – we cannot rightly conceive of them as separate, and they do not step out of experience to create internal representations of that world and then act upon those models. In the next section on Postphenomenology we will see this view's

---

[35] I'll discuss anti-representationlism again very shortly, but it's important to say immediately that though I'm very sympathetic to Chemero's broad project, I'm not wholly committed to anti-representationalism in all cases. That humans might be able to act without representations in some cases and require them in others seems a likely way of explaining both the impact of embodiment and environment on cognition, but also the distinctiveness of human intelligence.

compatibility with, in particular, Peter-Paul Verbeek's position on the role of tool use in human intention and perception.

First, however, I must raise a discontinuity with my own approach: whilst, as will continue to be apparent, I support the position that humans form dynamical cognising systems with the equipment that they deploy (i.e., artefacts not only play a role in the kind of cognition that occurs, but form a cognising system with the user), I cannot fully support the notion of an unproblematic direct access to the world. In this chapter, I have suggested that we only ever encounter things in changing composites that allow room for surprise, and in Chapter 4 I will offer further ontological support for this claim beyond cognition – we can't even *touch* things as they are. I don't believe, however, that this means that I have to throw my chips in with representational, computational, or non-dynamical approaches to cognition. The whole-composite objects that we form systems with for cognition and action (principally technological artefacts for our discussion here), objects that we access purely enough for Chemero's radical approach to hold, are *coherent* with reality, and they depend upon it, are partly a product of it – the gestalt objects that we encounter are parasitical off of the real world, only able to appear as they do because of the real object in the world, even as they are always partially misperceived. As Gibson's affordances are real properties of objects, but can only be understood with regards to some interacting creature, so I want to maintain that our contact with ever-changing gestalts can only be understood in terms of the real objects that they partially rely upon in order to give them their particular form. This discussion must rest for the moment; it will take time to outline out this claim more fully in Chapter 4. Here, it suffices to say that we have largely been discussing the radical position that Chemero advocates throughout – humans think and perceive together *with* their artefacts, not alongside them, and this is why the particularities of those objects can so dramatically impact upon our conceptions of the world.

3.4.4 Postphenomenology and entwined embodiment

This chapter has developed a theory of encountering objects, and particularly our technological artefacts, as malleable gestalts and the kinds of physical, psychological, and cultural forces that impact upon their formation. This fleshes out the definition of technology from Chapter 2 by saying that if the word "technology" best describes a type of encounter then *this* is the nature of what is encountered in that type of interaction. Part of developing expertise is training the whole-composites that we experience, and part of expert encounters is their only ever being partial – there are elements that will always be beyond our experience or control. The realities of objects and of our bodies, their hidden aspects and the insufficient gestalts that we form of them, leave plenty of room for surprise, for unexpected features of the world to manifest and interrupt action. In Chapter 4 we will explore the full implications of this capacity for surprise, but I want to finish here by positioning the work of this chapter, and of this book more broadly, as occupying, after Don Ihde, a "postphenomenological" stance, one that unites radical 4EDS cognition, embodiment, hermeneutics, technological artefacts, and the experience of practice.

A postphenomenological outlook offers a bridge between each of the fields explored in this work, sometimes explicitly, but sometimes by bonds of affinity that we need to explore further or to explore for the first time. For Ihde, postphenomenology is an updating of phenomenological concerns particularly with regard to the role of technology in human action and perception. The field tends to be case-study and praxis focussed, intently interested in how humans actually deploy artefacts (for Ihde technology is always physically instantiated), and combines elements of pragmatism and classical phenomenology to form an approach sensitive to embodiment, social forces, and the interactions between humans and their tools. In this regard, aside from the assertion that technology must be material, I hope that Ihde's interests already start to resonate here.

Ihde's work also allies with the intimacy of the technological encounter that I have tried to emphasise throughout: "By focussing upon embodiment and inter-relationality, I have shifted away from the separation of the machinic and the human to the interaction of the machinic and the human" (Ihde, *Embodied* 46). This sense of interrelation leads Ihde to document the ways in which humans are influenced by their technologies, and he develops a language for describing these deep relations. In *Technology and the Lifeworld* he outlines four types of encounter with technological artefacts: (i) Embodiment relationships – where humans bring artefacts "onboard" and treat them as a part of themselves, what I've termed "Incorporation." (ii) Hermeneutic relationships – where humans use artefacts to look "through," or to interpret the world, such as when recording readouts from a machine rather than directly observing a phenomenon, or bringing the universe or bacteria "closer" through a telescope or microscope (indeed making their observation possible at all). (iii) Alterity relationships – where humans consciously focus on an artefact as an embodied "other," for example drawing money out of an ATM machine where we must focus on the artefact itself in order to act.

And (iv) Background relationships – where artefacts don't occupy our attention or actions, but still flavour them by impacting on our environment, for example the presence of thermostatically controlled heating.[36] In establishing this language, Ihde investigates the ways in which the use of artefacts affects (and effects) perception and action.

A compelling example of how our perception might be altered by the use of artefacts is Ihde's investigation of astronomy. He speaks of Galileo's discoveries of the topology of the Moon, the phases of Venus, and the moons of Jupiter:

> These phenomena...had not been previously observed (perceived). But, they were not perceived directly by Galileo either, but through the mediation of an instrument, a technology. These were genuinely new perceptions, and perceptions which changed the world for the perceivers...To recognize that directed intentional human experience can embody a technology is a first step towards postphenomenology in that a material artifact can be taken into first person experience. (*Embodied* 57–58)

Here Ihde highlights classical phenomenology's placing importance on analysis from the first-person (indeed, phenomenology challenges the notion that there can somehow ever be analysis from "nowhere") and also the way in which artefacts mediate our encounters. Crucially, mediation and interruption, or insulation, are different things – technologies shape how and what we can perceive and how we can conceive of and put action into practice, but they don't separate us from the world in the way that some of the commentators from Chapter 1 feared. In Ihde's account, artefacts do not act as replacements for a more natural mode, they do something else: they represent an alternative or the only viable option.

Our artefacts can give us a new outlook on the world even as they change how we conceive of our place within it. Again, an astronomical example: "A hand-held telescope magnifies both the Moon and our bodily motion and thus makes it hard to maintain a fixed focus upon our observed heavenly body. Here is yet another clue to the complexity of embodiment: every change in our newly magnified world is also a change in our embodied experience" (Ihde, *Embodied* 58). To feel the slightest movement magnified a thousand- or ten thousand-fold, to

---

[36] For more on these relationships see Ihde's *Technology and the Lifeworld* (72–112) and *Postphenomenology and Technoscience* (42–44). Verbeek also discusses these relations (*What Things Do* 125–128).

make a minute gesture vaster than our natural satellite, is to feel distance profoundly. In this way, postphenomenology, as with my criteria of technological interactions, calls for us to think of technologies as more than just implements which exist outside of ourselves and to which we must turn in order to perform a task. Instead our artefacts alter the set of default practices that we consider ourselves able to achieve even in their current absence. We could not travel at speed without the car in the garage or the plane in the hangar; we could not type and print without the computer in the next room; we could not have hunted effectively without the spears in the rack nor butchered the catch without the knives near the hearth, and we cannot even *intend* to do so without these items. As Taylor describes this entailment: "it is not too much philosophy to say that the emergence of technology was and is intimately connected with the extension of the range of human intentionality... [T]he existence of [artefacts] not just allows actions but suggests them" (152).

What I would like to avoid, however, is the idea that postphenomenology is somehow deterministic. As Merritt Roe Smith and Leo Marx describe it, the worst instances of technological determinism suggest that "a complex event is made to seem the inescapable yet strikingly plausible result of a technological innovation" (xi). Carr, in a departure from the nuance he deploys in *The Shallows*, offers an example of how such determinism can make its way into mainstream discourse:

> When printed books first became popular, thanks to Gutenberg's press, you saw this great expansion of eloquence and experimentation... [a]ll of which came out of the fact that here was a technology that encouraged people to read deeply, with great concentration and focus. And as we move to the new technology of the screen... it has a very different effect, an almost opposite effect, and you will see a retreat from the sophistication and eloquence that characterized the printed page. ("How E-books Will Change Reading and Writing")

It is ludicrous to think that it was solely the affordances of Gutenberg's printed works, rather than an increasingly democratic access to knowledge, literacy, audience, and a raft of other cultural factors that gave rise to any increase in eloquence and experimentation. What a postphenomenological stance does suggest, however, is that artefacts, such as printed books, *will* have an effect *alongside* the host of other attendant factors; artefacts (and, I would argue, particularly those which we encounter as technological) are shapers and guiders, not determiners of cognition, perception, and thought.

Consider the alphabet. The alphabet is, primarily, equipment for preserving meaning and a technology, under my definition, for a great many people. If technologies were inert then the alphabet would have no effect on thought in and of itself; the capacities that it endows (writing, storage, etc.) might feed into a culture in a way that enables change, but the individual's use of the thing *itself* should not have a result of its own. But this doesn't seem to be the case. McLuhan, for instance, argued that the alphabet led to a certain kind of thinking:

> The alphabet is a construct of fragmented bits and parts which have no semantic meaning in themselves, and which must be strung together in a line, bead-like, and in a prescribed order. Its use fostered and encouraged the habit of perceiving all environments in visual and spatial terms – particularly in terms of a space and of a time that are uniform c, o, n, t, i, n, u, o, u, s, and c-o-n-n-e-c-t-e-d. The line, the continuum – this sentence is a prime example – became the organizing principle of life... "Rationality" and logic come to depend on the presentation of connected sequential facts or concepts. (*Massage* 44–45)

This equation of rational thought with linearity is something that we've already seen in the context of the codex in Chapter 1, and we also encountered Plato's views on the inception of the modern alphabet in ancient Greece and its effects on the early Greek literates in the last chapter. The first "complete" alphabet emerged in Greece, a system that included vowels for the first time, as opposed to the contextual cues of the earlier vowel-less Semitic scripts which required a thorough knowledge of how a passage would sound spoken aloud.[37] The Greek alphabet's completion of abstracting the spoken word into individually unintelligible written characters severed the alphabet from its relation to materiality, particularly in comparison to the world-imitative qualities of earlier (or contiguous in the case of, e.g., Chinese calligraphic script) pictographic writing. Ong suggests that "it does appear that the Greeks did something of major psychological importance when they developed the first alphabet complete with vowels. [Eric] Havelock[38]... believes that this crucial, more nearly total transformation of the word from sound to sight gave ancient Greek culture its intellectual ascendancy" (90). Such ascendancy derived from "the philosophical thinking Plato fought for

---

[37] For more on the origins of early alphabetical writing see John Man's *Alpha Beta*.

[38] See *Origins of Western Literacy*.

[which] depended entirely on writing...[F]ormal logic is the invention of Greek culture after it had interiorized the technology of alphabetic writing, and so made a permanent part of its noetic resources the kind of thinking that alphabetic writing made possible" (Ong 24 & 52). Despite the scepticism about writing as a tool of memory that we saw cyphered through Socrates, Plato's thought, similar to Derrida's writing in *Glas*, depends on the affordances, the material metaphors bound up, not in the matter, but in the form of the technology of the alphabet.[39]

It is precisely this "kind of thinking" that the alphabet can engender that John Gray attacks at length in his anti-techno-utopian *Straw Dogs*:

> The pictographs of Sumer were metaphors of sensuous realities. With the evolution of phonetic writing those links were severed...It is scarcely possible to imagine a philosophy such as Platonism emerging in an oral culture. It is equally difficult to imagine it in Sumeria. How could a world of bodiless Forms be represented in pictograms? How could abstract entities be represented as the ultimate realities in a mode of writing that still recalled the realm of the senses?...Plato's legacy to European thought was a trio of capital letters – the Good, the Beautiful, and the True. Wars have been fought and tyrannies established, cultures have been ravaged and peoples exterminated in the service of these abstractions. Europe owes much of its murderous history to errors of thinking engendered by the alphabet. (56–58)

And there, again, is the familiar claim: equipment, here the alphabet, places a barrier between its users and the true nature of the world. But if we instead take a postphenomenological stance, then we see human and alphabet as *intertwined*, with the equipment offering up an alternative or modified set of adumbrations to perceive, not a blocking of what continues to persist. To Ong and Havelock, the alphabet's mediation is exactly what caused Greek thought to be so powerful: temporary separation from the world was what was required in order to think abstractly. This is not to say that the alphabet was the sole arbiter of Greek thought, and of the Western philosophy which followed it, merely that it played a (typically unsung) role in its development – as it separated its thinkers

---

[39] For a discussion of the material metaphors bound up in print and e-reading see my doctoral thesis *Incorporating Technology*. The thesis includes early drafts for some of the ideas presented here, but with a greater focus on e-reading, experimental fiction, and critical theory.

from the world its existence as something manifested *in* that world continued to affect the thought of any thinker who incorporated it into their cognitive strategies. In short, the postphenomenological position holds that it is not deterministic to recognise the ways in which the things that we encounter most intimately, be they alphabets, hammers, or codices, have an effect on our cognition and the things that we can conceive of, and this is exactly the domain of technological use.[40]

Though we will always come back from truly technological engagements somehow changed (hence my criterion of Domestication), that change is produced by more than the affordances of the equipment (hence the criterion of Communality and the discussion of the range of participatory factors in the production of gestalts discussed throughout this chapter). To Gray, as for the resisters of technology that we encountered in Chapter 1, mediation seems to produce too many mistakes of thinking, but a postphenomenological stance further nixes the idea of some unmediated state from which we've somehow detoured. If nothing else, our bodies are the first devices that we technologise, marshalling their imperfections as we carve up the world. Our further incorporation of artefacts simply alters where and how we make the cuts.

Whilst various takes on the ways in which technology affects our actions and perceptions has been my primary concern throughout, the start of this chapter also considered the converse, how our perceptions and actions affect the nature of the artefacts that we encounter, and this gives us a circle. We perceive not actual things, but whole-composites that are deeply affected by culture, convention, experience, practice, and embodiment. And sometimes, often, our perception passes through (hermeneutic relation), or acts with (embodiment relation), an artefact and the properties of that thing shape what we encounter through or with it as well as altering our sense of our own embodiment which can then retrigger a shift in perception. This is an extension of the complexity of

---

[40] Ihde finds an illustrative example of such non-deterministic nudging in ancient writing tools:
   some actions performed through [them] are easier than others. For example, making straighter marks with either a cuniform stylus – or much later a Roman chisel making phonetic inscriptions upon stone – allows straight cuts more easily than curved ones, such that I, V, W, are easier than O, U, S. Then, too, if one has a long and complex text to produce, speed is called for and one can in many, many written languages detect how abbreviations, an evolution toward a simplification of script, takes place. The stylus does not determine this, but the relative ease or difficulty inclines the user to often take paths of least resistance. (*Embodied* 73)

embodiment that Ihde discusses: humans and artefacts become deeply *entangled*, continually affecting one another and mutually constructing the encounters that they are involved in. While we finish up here I will remain focussed on technology's impact on human thought and action, but the next two chapters will explore in much greater detail what this entanglement means both for the design of artefacts (a far from directly intentional process) and for our sense of the world.

The postphenomenological entanglement of humans and artefacts, I would like to suggest, naturally fits with Chemero's radical approach to Embodied Cognition and demonstrates, to my mind, a clear direction for current philosophy of technology: uniting postphenomenological investigations into the impacts of artefacts on human experience with the evidence of contemporary 4EDS Cognitive Science, and in the process continuing to radicalise each field. Chemero shows the path for radicalising Cognitive Science, and postphenomenological work could prove a fine source of support and further research questions in this area. And by bringing the work of Cognitive Science explicitly to bear on issues surrounding the entangled experience of humans and their artefacts we might also continue to radicalise postphenomenology in a vein already begun by Peter-Paul Verbeek.

Verbeek, in his fantastic work on postphenomenology and the philosophy of technology, *What Things Do*, aims for a radical postphenomenology in much the same way as Chemero's work on cognition is radical: he sees human beings and their world as co-constituting one another (113). But what is strange is that though both writers see the human and the world as being tightly coupled in perception and action, Chemero sees this as evidence of direct apprehension whereas Verbeek sees it as evidence of the *impossibility* of direct access, arguing that whilst phenomenology

> correctly pointed out that scientific disclosure of reality is not a disclosure of "reality itself" but always that of a quite specific kind...it failed to draw the conclusion that no final contact with "true reality" is possible at all...[It] is not to say that the world is only a construction, just that we can never know the world as it is in itself, but only as we disclose it. An uninterpreted world, a world in itself, cannot be experienced. (105 & 107)

I won't address this seeming impasse much further here, but in Chapter 4 I want to argue that they are, to some extent, both correct, though, for now, this chapter's prioritising of gestalts over direct perception clearly sees Verbeek winning out. By treating phenomenology not as a study of the world as-it-is, but instead as a study of the way in which humans relate to their environment, Verbeek aims to bridge an idealism which sees the world as only what is available to consciousness and a realism which stresses a wholly mind-independent realm:

> Postphenomenology can be viewed as an offshoot of phenomenology that is motivated by the postmodern aversion to context-independent truths and the desire to overcome the radical separation of subject and object, but that does not result in relativism. From the postphenomenological perspective, reality cannot be entirely reduced to interpretations, language games, or contexts. To do so would amount to affirming the dichotomy between subject and object, with the weight merely being shoved to the side of the subject. Reality arises in relations, as do the human beings who encounter it. Only in this sense is postphenomenology a relativistic philosophy – it finds its foundations in relations. (113)

This is why it is so important to stress embodiment: there is a real and impactful world of artefacts and human physicality that has effects on its own terms. What postphenomenology highlights, particularly in Verbeek's hands, and this is why I want to position my whole-composite approach here as postphenomenological, is the co-constitution of the *encountered* world (and its artefacts) and the particularities of human experience. But we can nuance this further.

3.5 Coda

I must disagree with Verbeek on one significant point, and one that sets the stage for the final move of this book: the postphenomenology that I'm interested in *does not* find its foundations in relations. The relationships between humans and their artefacts are clearly vital to my thinking, and postphenomenology completes the circle between the factors that impact human perception of artefacts and the impacts of those artefacts on human perception. But, for me, the foundation for this entanglement is actually a radical *division* – Verbeek's belief in our inability to access things as-they-are[41] is in fact at odds with his belief that what we encounter is, first and foremost, the product of relations.

---

[41] Verbeek discusses this inaccessible world as the existence of imperceptible "existences" (*What Things Do* 164) after Latour's "La Clef de Berlin" ("The Berlin Key"). Verbeek, as Latour, accepts that relations do not produce entities out of nowhere, but claims that those entities do not function as actants with essential properties until they enter into relations which structure what they are as actants. I want to emphasise the importance of all actants before they interact, what they both bring to and withhold from the table.

We have attended to the idealistic side of the equation long enough; in Chapter 4 I want to turn to realism, to an object-oriented ontology that I will argue provides an ontological base for these postphenomenological experiences. My aim will be to produce a radical posthuman stance, one which truly puts objects on a level footing with humans in our entangled interactions with one another whilst partially refuting Chemero's direct access and the primacy of Verbeek's relations.

4
Brushing Against Reality: Technological Interactions Require Knowledge

In the last chapter I explored what humans encounter when they use a technological artefact and the ways in which that encounter can be shaped by prior experience and embodiment as well as some further evidence for the ways in which technologies can shape other aspects of our cognition and perception. In some ways the argument of the next two chapters is simpler: the artefacts that we experience as technologies are embodiments of knowledge. And this might seem intuitively correct – the more that humans learn the better (for some value of "better") their technologies become. This is certainly true, but it's also not the whole story, or the only story, that I want to tell here about the relationship between technology and knowledge. Technology, particularly as viewed from a postphenomenological perspective, is one of the fundamental relationships that we can have with the world, and as such it necessarily opens up questions about knowledge and what it means to know. Much as with the word "technology," however, "knowledge" is a messy term which needs to be pinned down before it can be discussed, and because we are investigating material encounters, we also need to work out the relationship between the knowledge stored in things in the world and those who encounter that world – this requires a theory of both "information" and "data." I want to argue that not only does an understanding of technology require an understanding of such fundamental epistemological questions, but that the technological encounter that I've described can also give us an approach to how we come to know.

The approach to knowledge and knowing that I will take is fairly radically pragmatic: I want to argue that "knowledge" is any stored data that enables repeatable and successful action within an environment, that is, knowledge is "data *for*" ("information" will be positioned as "data *about*" or, more accurately, "data *because of*," stored aspects that can tell us something about a thing or an encounter – information is *readable*). What's most significant about this conception of knowledge is that it gives us an epistemological underpinning for our experience of expert technological use, but it also allows for artefacts to have their own epistemology – without resorting to suggesting that artefacts can cognise I do want to argue that they know, and that their knowledge stems from the same root mechanism as the more complex subset of knowledge that underpins human action.

I will address each of these themes in relation to technology whilst fleshing out the other side of the argument from the last chapter: what is it that we encounter when we use a technology? From the user's perspective, as we've seen, we meet a gestalt, a whole-composite, but it is important to note that these constructions are not the things themselves – it is to these real things that we must attend here; it's real things that know. The intended result of these final chapters, uniting the work of the rest of this book, is to have a coherent view of technological artefacts and human users as constantly in flux, mutually co-shaping one another in emergent and unpredictable ways during repeated engagements whilst always remaining both utterly real in themselves and yet encountered as filtered metaphors for, and abbreviations of, that reality in their encounter.

In order to make this case I want to take an object-oriented approach that primes the way for a non-anthropocentric understanding of knowledge. Largely drawing on the work of Graham Harman, I'll start by outlining some of his project for an Object-Oriented Ontology (OOO), how it might further our understanding of encountering whole-composite objects, and its usefulness for discussing technological artefacts. This chapter isn't about advocating for OOO as an overarching philosophical project, but about recognising the affinity between how Harman sees objects manifesting in perception, and enduring outside of relations, and the distinction between amateur and expert encounters with the same artefact outlined in the last two chapters. The most significant feature of OOO for our discussion here will be the assertion that we can never "satisfy" objects, never fully understand them as what they are, leaving all things with the capacity to surprise anyone (or anything...) that encounters them. Counter to Harman's claims, however, I do want to suggest that we can get *closer* to what things are in themselves, that their infinite retreat need not be compromised by our increasing comprehension of them (and, indeed, that this is what the process of technologising and developing expertise relies upon). It will, I admit, be a strange kind of comprehension.

With the object-oriented ontological underpinnings of the chapter established, I then want to briefly outline what I intend here by "knowledge," "data," and "information," and their relevance to discussions of successful action in the world (principally by human agents). In Chapter 5 I will build on these ontological and epistemological roots, pausing the discussion to introduce a new theme: that technologies can be seen as the products of evolutionary forces. This then enables us to deploy a branch of the field of evolutionary epistemology that sees animal bodies as embodiments of knowledge about their lived environment; this, I will argue, can then be extended out to discussions of our equally evolved technological artefacts. By the end of these chapters I hope to have a rich thesis for (a) why I believe that our expert interactions with artefacts that we might describe as "technological" are intimately related to our knowledgeable actions in the world more broadly and (b) how those technological artefacts might be said to know us in turn, with rapid change in adoption and use of equipment occurring in direct proportion to the mutuality of knowing (something of real significance for the debates around e-reading). In this way, I intend to unite the ontological, epistemological, and postphenomenological concerns of this book around its primary interests in the significance of user and artefact embodiment and encounter, further flattening out the hierarchy of subject and object and emphasising the role of artefacts in our experience. I hope, therefore, to address the call that Ian Bogost (another foundational figure for OOO) makes for stressing the importance of a posthuman outlook for object-oriented philosophy:

> Posthumanism...is not posthuman enough...[H]umans...possess a seemingly unique ability to agitate the world, or at least our corner of it (although this...is a particularly grandiose assumption, given that humans interact with only a tiny sliver of the universe). If we take seriously the idea that all objects recede interminably into themselves, then human perception becomes just one among many ways that objects might relate. To put things at the center of a new metaphysics also requires us to admit that they do not exist just for us. (*Alien* 8–9)

Bogost's *Alien Phenomenology* insists on the life of things independent of human perception, and OOO gives us a way of exploring the limits of our access to this life, as much as we can, even within the seemingly human-dependent and subject-prioritised realm of technological use. The rich life of real things is what ensures that technology can always push back, that the process of technologising can always be interrupted, and expertise is always provisional. OOO offers us the blueprints for every object's escape and thereby drags us deeper into the life of our technologies as the objects that they are before we even approach them in use.

4.1 OOO

Object-oriented ontology seeks to remedy an anthropocentric turn in Western Philosophy that sees objects as reduced to their correlation with perception by human observers. OOO traces this stance to the transcendental idealism developed by Immanuel Kant in his *Critique of Pure Reason*; for Kant, the thing itself is necessarily unknowable, and his *Critique* aims at unravelling the confusion between the available appearances of phenomena and the hidden reality of noumena, the distant things-themselves. The philosopher Levi Bryant, another significant proponent of the object-oriented approach, argues that

> [i]n beginning with the hypothesis that objects conform to mind rather than mind to objects, Kant who genuinely sought a secure grounding for knowledge and freedom from the endless debates of metaphysics, paradoxically rids us of the need to consult the world or objects. For as Kant himself observes, this shift or inversion allows us to discern how it is possible for something to be given in advance. Yet if the world is given in advance, then there is no longer any need to consult the world or objects... [P]hilosophy, at this point, becomes self-reflexive, interrogating not being or the world, but interrogating rather the mind that regards the world. ("Onticology – A Manifesto for Object-Oriented Ontology")

In this way, responding to the influence of Kant, OOO reiterates classical phenomenology's call for a return to "the things themselves." But as Bogost highlights above in his *Alien Phenomenology*, an object-oriented approach aims to try and deal with objects far beyond their reduction to human sense perception.

A point that each of these traditions share is Husserl's assertion (as discussed in the last chapter) that "nothing in perception is purely and adequately perceived," a claim that can be traced back at least as far as the Platonic forms. For Kant this renders noumenal reality wholly inaccessible, but Husserl's later phenomenology moved beyond this seemingly immovable opposition of phenomena and noumena. As Dan

Zahavi argues, phenomena for Husserl are "understood as the manifestation of the thing itself, and classical phenomenology is therefore a philosophical reflection on the way in which objects show themselves – how objects appear or manifest themselves – and on the conditions of possibility for this appearance" (*Husserl* 55). This resonates with the discussion of whole-composites and postphenomenology from Chapter 3: the reality of things, their real embodiment, affects the types of manifestations that we encounter in perception, even as that perception is shaped by, for example, a whole host of cultural and material forces. OOO becomes a wholly appropriate base for further thinking these ideas through, marking an odd comingling and refutation of Kantian scepticism and the Husserlian manifestation of reality through phenomena: similarly to Kant, for OOO we can never get to the things themselves, and yet objects do anything but conform to human minds;[1] similarly to Husserl, phenomena are seen as emanations of an ontological reality, and yet OOO maintains that the observation of such emanations doesn't bring us any closer to things as they are. As we'll go on to explore, these insights help in providing us with a coherent and explanatory ontology for postphenomenological experience while further emphasising the reality of our artefacts on their own terms rather than reducing them, as Verbeek does, to their relations with their human users.

To reiterate, and to state this clearly before the necessary complexity of the case to be made: for postphenomenology there may be a real world out there, but the way in which it manifests itself is created in the relationships that humans enter into with it – the world as it appears is a product of relations and cannot be conceived of outside of those relations; Kant's influence looms. For Chemero's radical approach to cognition we must access the real world directly, even if perception can sometimes be tricked – the world as it appears is a product of relations, but those relations must provide access to the real. For OOO, in turn, we can never access the real world directly, but the way that it appears to us is also not created solely in the relations that we enter into; the way that it appears to us could only ever be as it is because of what the real objects really are *before* their relations. A hammer will appear to us as a hammer-like-this when I am like-I-am in this moment (which is like-it-is) and because the

---

[1] As Harman puts it: "my position is in agreement with Kant's on the things-in-themselves. It's the other side of Kant (privilege of the human-world relation) which is more dominant these days, and on that point I have nothing to do with Kant" ("Quick Response to Shaviro"). It is in this denial of privilege to the human-world relation that OOO significantly differs from both phenomenology and postphenomenology.

hammer is what-it-is. The *meeting* doesn't create the appearance of the hammer – though it could appear in an infinite variety of other ways, it could appear in no other way *right now to me* (and the meeting certainly doesn't affect the real thing in the world which has, as Chemero describes, its real affordances whether they are encountered or not). For Verbeek and postphenomenology it is the relations that come first, that is where the emphasis lies – we cannot access the world so we must consider the relations which construct the actants that enter into them. For OOO, and for me, the *objects* must come first – we cannot access the world directly, but we must consider that the relations are pregiven *because of the nature of the objects that are doing the relating.* The objects don't appear in this way because of their relations; the relations are possible as they are because of the nature of the objects. This primacy of objects is what gives the space for surprise – as we will go on to explore, all things, even only as they are given as whole-composites in relations, are more than the sum of their interactions; there is always more to give, but also always more to hide away. I want to explore how we could possibly get closer to such things and to argue that this is what the process of technologising must require – expertise gives us a model for approaching the world.

Part of my goal here, then, is to take some of OOO's principle assertions and demonstrate their potential usefulness for thinking about artefacts, to argue for OOO's place in both the philosophy of technology and also Cognitive Science where it might offer further support to Chemero's radical approach whilst also offering a posthuman focus on the material world's reality and effects independent of perception.[2] The ideas from OOO that I principally want to work with are (roughly in order of importance to my argument):

- Objects can never be wholly known.
- All objects (regardless of their sentience) interact with one another in a manner intimately related to the typically privileged human-object interaction.
- Objects cannot be reduced to the sum of their components (and components are always objects too, with their own realities and often their own equally real components).
- Objects exist prior to their relations with other objects.

---

[2] And also challenging faith in direct perception without resorting to mental representations – this won't be a significant focus of this chapter, but I hope to show that no encounter can ever be seen as direct, even when unmediated by technology or perception. As we will see, any kind of contact, under OOO, is always a reduction.

In the early stages of this chapter I ask that the reader bears with me during a couple of lengthy quotations – because it isn't my aim to rewrite OOO, but instead to draw upon it, I intend to let people speak for themselves, not least because the newness of the field, and particularly its discussion outside of Philosophy, resists glossing. As OOO's originator and leading proponent, I will predominantly deal with Graham Harman's work here; his focus on the role of objects and the importance of Husserl's phenomenological insights are also closely allied with many of my own concerns.

4.1.1 Breaking the tool-analysis

In *Tool-Being*, the first significant work in the OOO canon, Harman interprets Heidegger's tool-analysis as holding "*the whole of the Heideggerian philosophy*, fully encompassing all of its key insights as well as the most promising of the paths that lead beyond them" (15, emphasis in original). Against the more common readings of Heidegger that we've already discussed, Harman claims "that the tool-analysis is neither a theory of...human praxis, nor a phenomenology of a small number of useful devices called 'tools.' Instead Heidegger's account of equipment gives birth to an ontology of *objects themselves*" (1, emphasis in original). For Harman, Heidegger uses the hammer's melting away during use as a front for a wider ontology of all objects, the full implications of which could not be accepted even by Heidegger himself who remained distractingly fixated on the human as the site of cause and ontological priority.

Taking two terms that have become familiar to us, readiness-to-hand (tools invisible in use) and presentness-at-hand (tools merely observed or somehow faulty and thereby intruding into consciousness), Harman describes an inner essence for objects which will always retreat from the human, and in fact from all other objects that they might encounter: "readiness-to-hand (Zuhandenheit) refers to objects insofar as they withdraw from human view into a dark subterranean reality that never becomes present to practical action anymore than it does to theoretical awareness" (1). In this way, readiness-to-hand, for Harman, doesn't describe phenomenological experience, but instead an aspect of all objects' ontological reality; presentness-at-hand, in turn, doesn't just describe the faulty or merely observed, but rather the *encountered* in general.[3] Harman asks us to look beyond the language of the tool-analysis to its implications: "Whatever Heidegger's intentions may have

---

[3] This distinction is further described in *Tool-Being* (44–49).

been, his theory of equipment applies to all entities...It is vital we not be misled by the usual connotations of the word 'tool'" (2).

The central argument of *Tool-Being* is that the ready-to-hand, the invisible life of every thing (a broadening of what I referred to in the last chapter as the "escapological" tendency of our artefacts), means that no object is ever encounterable as what-it-is. Influenced by Xavier Zubiri, Harman is interested in the non-relational aspects of objects, the essences that recede from and precede all relations with anything in the world.[4] That which we do encounter, the present-at-hand aspects of our contemplation, the phenomena that Husserl saw as emanations of a fundamental reality, are something else entirely.

The mainstream reading of Heidegger's tool analysis sees readiness-to-hand as the most intimate interaction with an object's essence,[5] whereas presentness-at-hand represents the greatest separation where we disengage and merely contemplate an outer surface. For Harman, however, the seemingly "intimate" interaction between user and incorporated tool and the "distant" interactions of mere observation are of the same order: they both reduce an object to a set of qualities (e.g., grip-able, heavy, red, an extension of the arm for use) that do not bring us closer to the thing itself. They, the grasping and the observing, are both forms of presentness-at-hand, calling a new object into being (what we will come to call a "sensual object") whose existence is some combination of encounterer and encountered.[6] If readiness-to-hand is about the essence of a thing then it necessarily describes what is independent of us, and what is independent of perception is, as Kant recognised, unencounterable no matter how seemingly intimate the interaction:

> If I stare at a bridge [presentness-at-hand], this bridge-appearance is a parasite off of my Dasein, and could hardly be less independent. If not for me, this appearance could not exist. What is truly independent of

---

[4] Harman cites Zubiri's *On Essence* in this regard. Unlike Zubiri, however, Harman doesn't believe that only certain things can possess such an essence; in fact everything, and every combination of things, is an object with an essence and qualities that recede from any encounter.

[5] Though, as I'll reiterate shortly, I find the mainstream reading of the tool-analysis useful, I haven't (yet) argued here that readiness-to-hand as invisible use draws us closer to either the world or the object, simply that it is a sign of expertise. It is around this point that a tension will form between my discussion of expertise and Harman's object-oriented approach.

[6] And here is our first connection back to Gibson/Chemero's affordances and Verbeek's co-constitution – sensual objects are manifestations that only exist in relations and yet have real effects on those who encounter them.

me is not the *occurrent* bridge, but the *executant* bridge, the bridge that is hard at work in enacting its own reality and all that this entails." (Harman, *Tool-Being* 125, emphasis in original)

The occurrent bridge is not born of the executant bridge alone (as we saw with the influence of culture and embodiment on the production of gestalts), and the essential executant bridge acts independently of and is inaccessible to the observer (some aspects always escape). Where Husserl saw the occurrent object as giving us a way towards encountering the executant object, and where the mainstream reading of Heidegger sees invisible use (seemingly the bypassing of the occurrent) as possessing the same potential,[7] Harman instead maintains that there is *no* way to experience an object's essential character, directly or indirectly. In this way, he sees Heidegger's tool/broken tool, ready-to-hand/present-at-hand, divide not as describing an ontic distinction between two kinds of encounter as it does in the standard analysis, but instead as describing two sides present in every thing (45): the parasitical and reductive occurrence of use or observation and the escapological executant essence which evades all such attempts at contact. This is all about a commitment to reality that keeps on keeping on even when we close our eyes, fall asleep, or tie our hands behind our backs: "Whether my hands or my intellect alerts me to the electrical conductivity of copper, neither sensing nor knowing is what conducts electricity throughout the world" (Harman, *Quadruple* 28).

I want to go on to challenge Harman's position, at least slightly – as might be expected from the themes of this work, I will argue that there *is* something special about use over contemplation that I don't think Harman accounts for, and that *does* draw us closer to the essential life of things, and that this move is at the heart of expertise. I also won't adopt Harman's distinction of tool/broken tool, ready-to-hand/present-at-hand here; those terms, in the conventional reading of Heidegger, still have too great a usefulness, particularly to my own argument. As we've seen in previous chapters, readiness-to-hand as invisible Incorporation during use is a real phenomenological and cognitive feature of expert interaction, and as such I have not, and do not evoke Harman's reading when I use those terms. Instead of ready-to-hand/tool-being I will use the more conventional terms "essence" or "real" to describe the ever-escaping being of the thing in itself, and "phenomena" or, a term from Harman, "sensual" to describe the (gestalt) objects that appear to our perception.

---

[7] Ditto Chemero's direct access.

4.1.2 Flat and non-relational

Harman's radical re-reading of Heidegger's tool analysis sets the stage for the project of OOO and its flat, non-relational ontology. Before I go on to talk about how Harman sees the structure of objects encountering one another in the world, and the implications of OOO for our discussion of technology, it is important to establish what is meant by "flat" and "non-relational," perhaps the most controversial of OOO's features.

4.1.2.1 *Flat*

Harman's ontology is flat in that it describes all objects' equal status as objects, and this includes objects held in cognition independent of physical realisation, for example, a remembered childhood home, a predicted outfit on a friend, or a mythological basilisk. This flatness bears a similarity to actor-network theory's (ANT) agnosticism to the sentience of the actors that it investigates (Bruno Latour preferring the term "actant" for this reason). ANT's actants each have agency, and each affects the network that they are involved in, whether they are, for example, scientists or their instruments, policemen, drivers, or speed bumps. For ANT, actants don't have equal agency, but they are all equally (flatly) actants. Harman similarly contends that he has "never held that all objects are 'equally real.' For it is false that dragons have an autonomous reality in the same manner as a telephone pole. My point is not that all objects are equally real, but that they are all equally objects" (*Quadruple* 5). The implications of this for our discussion here, then, will be based around the idea that an encountered whole-composite e-reader, despite its reductive and malleable nature as a truth only in our perception, is as much an object as the real artefact which always escapes our every attempt to perceive it as-it-is. Husserl's phenomena, Gibson/Chemero's sets of affordances, and Verbeek's encountered artefacts are each equally objects, but I do not believe that they are as equally real as the objects that lay behind them (and nor do they prevent those real objects' escape). That these objects, as with any sensual phenomena, are objects in their own right is evidenced by their real qualities and real effects which occur at least partly independently, certainly distinctively, of the effects of the real retreating object in the world.

4.1.2.2 *Non-relational*

The non-relational aspect of OOO demands a move away from most readings of Latour, ANT, and the Process Philosophy of, for example, Alfred

---

[8] See, for example, Whitehead's *Process and Reality*.

North Whitehead[8] (each of which have influenced both Verbeek and Chemero). The most important work of these writers and disciplines has been in emphasising the centrality of dynamic relationships, networks, and the equal status of humans and non-humans as actants with agency in the world; this has been a significant step for a great many fields and has been essential in recognising the complexity of the lived world and the relations that occur within it. I certainly side, as I hope is by now clear, with a view that embraces the agency of non-human actants and the importance of the relationships that form between (in particular) artefacts and their users, but OOO breaks ranks, unfashionably, with the notion that such actants are *constituted* by the relations in which they find themselves. Again, real objects of all kinds precede the relationships that they enter into and are always more than the sum of those relations. As Bogost describes this break:

> OOO is not the same as process philosophy or actor-network theory. For Whitehead, entities do not persist but continuously give way to one another – metaphysics amounts to change, dynamism, and flux...The successive actual occasions of experience amount to an undermining of objects into more basic components that perish immediately...Latour allows for the uncontroversial existence of things at all scales. But in the networks of actor-network theory, things remain in motion far more than they do at rest. As a result, entities are de-emphasized in favour of their couplings and decouplings. Alliances take center stage, and things move to the wings. (6–7)

OOO, then, can be characterised as possessing a faith in things on their own terms independent of their encounter with any other entities. In this way OOO runs counter, on the one hand, to a scientific naturalism which searches for reality at the smallest possible units (cells, atoms, quarks, super strings...),[9] and on the other to a belief in a fundamental

---

[9] Harman describes this view as "undermining" objects, seeing all the action as occurring at a lower level and the object being either a naïve product of imagination or a simple sum of its parts. Unlike Harman, however (who argues for objects as being necessarily made up of infinite constituent parts, rather than being comprised of fundamental smallest units, in order to retain their illusive natures (*Quadruple* 112–113)), I have little problem with a final plane of quarks or strings that are indivisible. We can be realists and speak of quantum effects without undermining objects; to consider electrons as objects and as affecting the objects that they make up is not to be inherently undermining as long as we accept that objects have a reality that cannot be reduced to a description of their physical makeup by scientific best practice. We can accept that things are more than the

flux which sees objects as necessarily subordinate to the relations in which they take part.[10]

Harman's challenge to an undermining scientific naturalism, or materialism, recognises the effects that objects have independently of their component elements. This is important for us here in two ways: (i) because real (hidden) objects have real effects as themselves – the escapological tendencies of artefacts, for example e-readers, mean that we can always be surprised by an aspect that we don't encounter, but we are surprised by an effect of the e-reader *as what it is* ("I didn't know it could do that!"), not simply from some more fundamental component (e.g., its atomic structure). The e-reader has a reality as an object, the effects of which emanate from a real thing that exceeds the sum of its parts. And (ii) gestalts have, as we've already seen, real effects produced by what they are – the gestalt e-reader affects what I can conceive of doing in a real way, even if it itself is a misprehension of the materiality and capacities of the real thing. As Harman puts it, a scientific naturalism that constantly looks for reality, and real effects, at a deeper level

> discounts the possibility of larger emergent entities. Even if all nations of the European Union are composed of quarks and electrons, we

---

sum of their parts; what harm does it do to know the fundamental parts? The object's reality cannot be reduced to this mere accumulation. Even if we know the exact nature of each piece, Heisenberg's Uncertainty Principle (at least) offers a final impasse for certainty – the capacity for surprise is always maintained. To know everything about, or to even posit a total knowledge of everything about a fundamental particle or string is not to know (or to suggest the possibility of knowing) everything about the objects that they comprise. That objects surprise in different ways to the sum of their components supports Harman's position; there may be fundamental particles, quarks or strings that we can divide no further and that make up all things (I incline towards monism in this regard as it seems the way that science points – at the bottom of everything there will come the indivisible, but perhaps this base is multiple, like the "colours" of quarks), but this doesn't mean that the essence of all things can be reduced to or explained by this fundament. Far from it; in fact this fundament can be changed in ways that, at its scale, seem dramatic while the object persists in its full richness, to be encountered as what it is and to be what it is. In short, super-strings, or whatever "deeper" layer comes next, needn't produce monotony – everything physical may be united in being made up of one or a few fundamental things, and yet not be one, not just be mass. It is remarkable and weird enough that a true object-oriented ontology could sit on such a strange layer, a reality of real complex objects based upon, but not satisfied by, another secret with no mystery.

[10] What Harman terms "overmining" – all objects are subordinate to, produced by, or illusions of dynamic relations.

can shift these particles around to some extent without changing the Union. There are countless different numbers and arrangements of particles that could be tried without the Union itself being changed...which suggests that the object is something over and above its more primitive elements. (*Quadruple* 16)

What is true of the union is as equally true for the hammer, the e-reader, and of any other object in the world.

It is essential to keep an emphasis on the co-constitution of phenomena (such as expert technological use) whilst recognising that such relations do not constitute, for example, the user or the artefact as essential objects, that is, they retain their escapological potential as there are always features of every object that precede and evade their relations in the world. In this regard, it's worth quoting from a conversation between Harman and Latour at the London School of Economics captured in *The Prince and the Wolf*.[11] Harman describes an increasingly orthodox reading of Latour's sense of objects:

> The...thing about actors for Latour is that there is no hidden essence or potential in them...An actor simply is what it is, which means that an actor contains all of its qualities, or contains all of its relations with other actors. There is nothing hiding behind those qualities and relations. An actor is wholly deployed in the world in every second. There's no cryptic reservoir hiding behind what the thing is doing here and now, what qualities it has here and now. The reality of the actor is its way of perturbing, transforming, and jostling other things...Now here's the paradox[:]...His whole philosophy is about relations: the way that things interact, the way they form networks, alliances, and relations. And yet, since everything happens in one time and one place only, and every actor is utterly concrete, this means that actors are completely cut off from each other as well...[An actor] can't possibly endure from one instant to the next because it's so utterly concrete that even the smallest change essentially makes it a new actor: unless some other actor does work to establish that the change wasn't important and it's actually still the same thing. But another actor is required to do that." (Harman, "The Transcript" 26–29)

---

[11] The book's pleasing title stems from Latour describing the philosophers that "chase" him as a pack of wolves, and Harman's book on Latour being called *The Prince of Networks*. The LSE dialogue saw the meeting of Prince and Wolf.

For Harman this is untenable. For change to be possible an object (or actor) cannot be "exhausted" by what it is currently doing, by what it is currently related to – there must be something left over, out of contact, that can be drawn on in order to produce change whilst retaining what a thing is as itself; objects resist being used up in each moment. When Verbeek suggests, as at the end of the last chapter, that "[r]eality arises in relations, as do the human beings who encounter it" then we cannot take this to mean that what it is to be real or to be human is to *firstly* enter into relationships with the world. To accept this would be to give in to the same problem plaguing Ireneo Funes: I, with the e-reader in my hand, am a wholly different actor than when I hold the printed book. Whilst the relationship that forms between me and the artefact absolutely has different potentials than myself or the artefact alone, OOO's argument is that we also must have realities as what we are before entering into those relationships, and though I might be shaped (Extended, Domesticated) by my activities, those activities do not constitute me or encounter every aspect of me – even as I form a part of some new machine for acting, aspects of user and tool always escape their meeting.

I won't dwell on this further here except to say two things: firstly, the important point for my argument is that regardless of the intimacy of the interactions between objects in a network, such as in an act of expert use of a technological artefact, there will *always* be some features of those objects that retreat away from contact – no set of relations satisfies what the object is, let alone determines what it is. Secondly, Latour himself, in the same conversation, seems to agree:

> [Y]ou [Harman] say that I associate myself with the doctrine that 'everything is relational.' And that I don't get, I simply don't get...I don't understand what it means to say that a thing is just its relations...And I don't understand why it is a problem in your book [the then unpublished *Prince of Networks*], since you do such a beautiful job of showing that from the very beginning the principle of irreduction is precisely to say that it's always at a cost[12] that you make a relation possible. ("The Transcript" 43)

For Latour, objects, actors, cannot be perfectly translated to new conditions, to new positions in space and time, and this is evidence of their

---

[12] This cost is essentially a reducing of the richness of a thing – every aspect cannot be encountered, something is always lost.

singularity (e.g., me with the e-reader vs. me with the book – that something is different is evidence of my real status prior to my relations). Harman (in Latour's view) misreads this as a claim that actors are the sum of their relations. We might see, instead, Harman as simply saying that objects are hardier than Latour (or Funes) give them credit for, that a cup in a kitchen at 9 am is the same cup with the same essence at 10 pm in a carpark – nothing is lost and *this* is evidence of the cup's singularity; in Harman's reading of Latour, for Latour this is not the same cup. To my mind Latour is somewhat conflicted on this ontological point, but I agree with him that the cup *as an actant* certainly has different effects on those that encounter it in its different settings, it's just that as an actant it must, by definition, be encountered, and encounters, as we will see, always produce a new phenomenal or sensual (and, in our discussion of artefacts, gestalt) object with its own effects – the real cup, however, remains.

4.1.2.3 Flat and non-relational and weird

A final point about OOO's objects leads us into the next section: One of the most radical of Harman's formulations is that the structure of encounters that he outlines doesn't just apply to human-oriented interactions. As an object-oriented philosopher, Harman extends his analysis to how *any* two objects meet, whether that be the object-human meeting the object-codex, the object-codex meeting the object-human, or the object-codex meeting the object-table. What this means, therefore, is that no two objects, whatever they may be and however cognisant they are of the interaction's occurrence, no two objects ever meet each other as they are, but only as they appear to one another: "No object ever unlocks the entirety of a second object, ever translates it completely literally into its own native tongue... [;] any object reacts to some features rather than others – cutting... rich actuality down to size, reducing it to that relatively minimal scope of reality that is of significance to it" (Harman, *Tool-Being* 223). This initially seems profoundly strange, but the point is (relatively) simple – an object must always encounter another object at least partially on its own terms and yet without satisfying it. When I hold a hammer it is clearly a hammer and not a sheep or bowling ball, but this doesn't mean that I "exhaust" the hammer, experience everything about it – as with our discussion of the production of whole-composites, I encounter a reduced set of aspects of the hammer that are shaped by a variety of factors. Similarly (ontologically identically), when the hammer encounters the nail it encounters a subset of qualities that emanate from the real nail, but again (clearly) it doesn't encounter all of them – the hammer's encountering the nail is as similarly reductive as our encountering the hammer. We can see this most vividly when treating ourselves as more passive objects: "If we unconsciously stand on a floor that has not yet broken, this standing relies on just a handful of qualities of the floor...Our use of the floor...makes no contact with the abundance of extra qualities that dogs or mosquitos might be able to detect" (Harman, *Quadruple* 42–43). A cup placed on that same floor makes a similarly reduced contact; "[t]hings oversimplify each other just as much as we do. It's not a special property of human consciousness to distort the world" (Harman, "The Transcript" 36–37).

We will return to this issue of object-object relations as part of the posthuman stance towards artefacts that I want to develop here, but our predominant concern will continue to be with user-artefact encounters. It's important to note from the outset, however, that these are in some ways much the same thing. I'll now go on to briefly flesh out Harman's ontology of every encounter before considering its implications for specifically technological interactions.

4.1.3 Fourfold encounters

As discussed above, Harman sees Heidegger's division of presentness-at- and readiness-to-hand as two aspects of every object – the simplification (or misperception) that we encounter and the inconceivable and ever-escaping essence. He further subdivides the encounter between all objects into a "fourfold," four aspects of the world and the significant "tensions" between those aspects. I won't spend too long outlining this division, but I do want to shore up the ontological assertion that aspects of all objects will always retreat from one another and explore the distinction between real and sensual objects, between, in my terms, the real artefact and the whole-composite that we encounter – I want to position this issue at the heart of understanding the nature of use by an expert who experiences a different object to that of the initiate.

The fourfold structure of objects, for Harman, begins with the division between the real and the sensual – the real object is only ever encountered obliquely and can never be fully known, while the sensual object is always only ever encountered (and it is, in large part, what I have been principally discussing throughout this work). In establishing the nature of the sensual object, Harman turns to classical phenomenology:

> Husserl's primary insight is the distinction between a thing and its qualities...If you circle a building, you keep thinking of it as the same building even though you're seeing utterly different qualities

in each instant, which means that there a [sic] distinction between the building and the qualities through which it is manifested. Those qualities are almost accidental; you just need to be seeing some qualities of the building in order to be able to see it. But you can keep circling the building and it stays the same. It remains identical, even if it's not real: even if it's the Tooth Fairy that you're circling in your mind, the same thing happens... And for one of his followers, Merleau-Ponty, there are no qualities independent of the thing. If you're looking at ink, a shirt, a flag, the black is different in each of those cases even if it is technically the same shade of black, because it's now impregnated with the underlying object, which is never fully present to you.[13] ("The Transcript" 38–39)

Here Harman draws on Husserl and Merleau-Ponty to establish that the encountered sensual object, the phenomena, is an object in its own right with its own qualities. This is part of why I used the term "whole-composite" to describe the objects that we encounter – things have a wholeness in perception (and reality) even as their adumbrations or our anticipations of them change. In this way, sensual and real objects are unified in their possession of both real and sensual qualities; qualities are features of objects. As we've already seen, objects, for Harman, are more than bundles of qualities,[14] more than the sum of their aspects, but there is also no such thing as a quality independent of an object, for example, there is no such thing as "roughness" in the abstract. (And this links us back to Chemero's reading of Gibson's affordances – an affordance must be born of the reality of the object that affords as well as the reality of the creature being afforded. I argue, however, that a creature must always be able to misrecognise an affordance because an affordance is also always more than the product of its appearance to a creature.)

These, then, are the four aspects of the fourfold: real objects, real qualities, sensual objects, and sensual qualities. They are best explicated

---

[13] See Merleau-Ponty *The Phenomenology of Perception* (e.g., 313). This aspect of Merleau-Ponty's work is closely allied with my conception of the construction of perceptual gestalts presented in the last chapter – reception matters.

[14] For example, an apple is tart, red, shiny, made of atoms, mostly carbon and water, etc. etc. etc. But to list these qualities, however exhaustively, is not to know the real object – the apple has these qualities, but it is not these qualities, nor any greater sum of them; the apple is a real thing that offers these features, at least in part, because of what it is.

through Harman's emphasis on four fundamental "tensions" between object and quality.

4.1.3.1 Real objects/real qualities (Essence)

Harman names this tension "Essence" and sees it as being linked to Gottfried Wilhelm Leibniz's monads,[15] the unified object as it is and the qualities that occur because of what it is.[16] "[T]here is a tension between the unity of a thing and the numerous central features that it unifies. This tension between a unified system and a plurality of traits is what we mean by the essence of a thing" (Harman "Intentional Objects for Non-Humans"). What is most significant here, particularly against some readings of ANT, is that an object's real qualities exist independently of its relations – real qualities do not have to be encountered to be real, and the real object is not what it is because of its qualities; it has those qualities because of what it is.

> *"Essence" describes the full nature of the real e-reader artefact that we will never wholly encounter and the qualities that it has because of what it is as a real object.*

4.1.3.2 Real objects/sensual qualities (Space)

Harman names this tension "Space" – "[t]hings make contact in space, but space also has distinct regions in which things can hide from each other" ("Intentional Objects for Non-Humans"). This tension is most closely allied with Heidegger's tool-analysis: real objects are only encountered via their sensual qualities, but these don't exhaust the object or represent a genuine contact; there is always a distortion, a relationship that Harman describes as an "allusion." In moments of breakdown, of the transition from (conventional) readiness-to- to presentness-at-hand, at moments of surprise, the real object is drawn to our attention, alluded to, without ever being revealed – "[t]he sensual qualities of the hammer no longer just swirl around the phenomenal hammer in the mind, but seem to be enslaved to a dark and hidden object that forever eludes

---

[15] "The Monad, of which we shall here speak, is nothing but a simple substance, which enters into compounds. By 'simple' is meant 'without parts'" (Leibniz, "Monadology" 1).

[16] "As Leibniz observed, even the simple unified monads must have diverse qualities: otherwise they would be interchangeable, with hammers equally able to function as drills, kidneys, dolphins, or monkeys depending on the whim of the observer" (Harman, "The Road to Objects" 174). Harman cites Leibniz's "Monadology" in this regard.

our grasp despite its apparently obtrusive malfunction" ("The Road to Objects"). We realise that the sensual features that we've been experiencing aren't simply bound to the object that we've been intending (a sensual object that we presume satisfies the thing), but instead that they are bound to a thing that we do not know and have not encountered, and this can shock us. Space is therefore the tension "between a real object and its relations, since relating to a thing only gives us a specific range of tangible qualities rather than the thing itself... Things make contact along specific surfaces but are not exhausted by this contact, and recede partially into private depths" (*Prince of Networks* 218). Things escape despite briefly seeming to be caged by perception.

> *"Space" describes both the expert invisible use of the real e-reader that reduces it to a set of encounterable qualities and the moments where the e-reader fails us, surprises us, returns to our attention and has its unknown real being alluded to, when we realise that there is more to know. How is it both ready-to- and present-at-hand in this way? Because, when we return to the mainstream reading, invisible use is still a reductive encounter, not getting to the thing's essence, and a surprising return to attention can be a revelation that such a reduction has occurred. Space is the tension of limited contact and the potential for the revelation of surprise.*

4.1.3.3 Sensual objects/real qualities (Eidos)

Harman names this tension "Eidos" for its connection to Husserl's eidetic reduction.[17] Eidos, here, describes the persistent and unchanging

---

[17] Husserl used *eidos* "to mean the subject of the set of predicates which could not be removed from a thing after having submitted it to a process of imaginative variation in short, the essence of a thing" (Novak 5). "Idea" was used interchangeably, in the Platonic tradition that Husserl evoked, to mean "form," but Husserl distances himself from "idea" due to the weight placed on the term by Kant, opting instead for *wesen* (essence/character/being/creature) or *eidos*. For Plato the unavailable Forms were what lay beyond every object in perception: "Plato names this aspect in which what presences shows what it is, *eidos*. To have seen this aspect, *eidenai*, is to know" (Heidegger, "The Question" 163). For Aristotle it was the immanent essence in all things: "ειδος δε το τι ην ειναι εκαστου και την πρωτην ουσιαν / By *eidos* I mean the essence of each thing and its primary substance" (*Metaphysics* 1032b1–2, ctd in Novak 1). For Husserl, phenomenology is similarly an "eidetic science," a seeking of what makes things what they are. The "eidetic variation" he advocates for reaching the essence of a thing is a thought experiment where aspects of the object are subtracted until it ceases to be identifiable as that object; all that is left, that "set of predicates" at the last point before it ceases to be the thing under consideration, is what makes it what it is, its *eidos*.

character of the occurrent object. The real qualities stem from (but do not comprise) the sensual object because of what it is – as with real objects' real qualities, a sensual object's real qualities are a fact of its being itself. In short, there is something about the sensual object of my intention that guarantees that it *is* what I encounter regardless of the manner in which I encounter it (e.g., in brightness or dimness; from the left or right; on a Monday when angry or on a Thursday when ecstatic; in ignorance or when informed) – Eidos is the whole-composite thing.

> *"Eidos" describes what makes the whole-composite e-reader that I experience consistently an e-reader (specifically the same e-reader) despite any change in its adumbrations, my perceptual abilities, or the milieu of the encounter. Surprise is the revelation of our mistaking Eidos for Essence.*

4.1.3.4 Sensual objects/sensual qualities (Time)

Harman names this tension "Time" with its philosophical antecedent being Husserl's adumbrations. The sensual object appears to us via a set of sensual qualities that emanate from, but do not express the essence of the thing; they are mere aspects (and they can fluctuate, giving us the perception of time passing). It is the sensual object that we encounter; its sensual qualities are only derivative of it. Sensual objects therefore "always appear in more specific fashion than is necessary, frosted over with accidental features that can be removed without the object itself changing identity for us" (Harman, Quadruple 21). We might describe this tension as what makes an occurrent object what it is *to us*, to the individual encounterer in a particular context; the qualities partially stem from distortions due to the limits and affectability of perception. These qualities are not, however, simply "made up":

> the various qualities of a hammer do not emanate only from the sensual hammer that I have in view. They also emanate from the real hammer that withdraws into subterranean depths beyond all access. Sensual qualities serve two masters, like moons orbiting two planets at the same time: one visible and the other invisible. (Harman, Quadruple 77)

The sensual qualities that occur are a product of both the essential object and the phenomenal, of the reality of the thing and the distorting effects of any and every encounter.

> *"Time" describes the manifestation of the whole-composite e-reader as I encounter it in a moment, the crust of qualities that emerge from the artefact*

because of what it is, but that iterate and mutate without impacting upon its essential is-ness, on the real or sensual, on the executant or occurrent.

4.1.3.5 The fourfold and technological interactions

How does this fourfold structure of encounter assist my argument here? Principally, it accounts for both the escapological tendencies of objects (real objects always withdraw, perception never gets to everything) and their manifestation in perception as something whole and always complete (sensual objects have real qualities which consistently maintain them as a gestalt unity). This, as we will continue to explore, underpins the potential for surprise that exists even in expert use (and gives us pause as to what it means to "know"). It also offers a deeper understanding of technological encounters when coupled with the (post) phenomenological roots established throughout this work coupled with an empirical faith that there are real things with real effects in a real world. It is particularly well allied with a postphenomenology that sees an encountered artefact as being co-constituted by interactions with a human user – the sensual object is just such a product of relations. But we must modify the postphenomenological stance to put real objects *before* the relations that they enter into; real things have real effects, a subset of which include influencing the whole-composites that appear in the relationship of human perception. I wholly agree with Verbeek that humans and artefacts intertwine and affect one another, but each intertwines with a sensual approximation, not some real thing – there are myriad factors, discussed in Chapter 3, that produce gestalt sensual artefacts, and it is these sensual objects that predominantly affect human intentionality. But real objects independent of sensual encounter also must have real and surprising effects, and this is the very opposite of human intertwining and co-constitution. Even during expert Incorporation, a mark of the technological, it is only ever a sensual object that we encounter as part of ourselves – the real object continues to escape.

OOO, as an underlying concern, is also useful because it gets us to focus on the bodies of things; it is via their bodies that every real thing attempts to encounter and divide up the world. Again, it is important to emphasise that such encounters are not just subject-object, but also object-object, a point that will underscore the increasingly posthuman stance of this chapter. Harman uses the example of tectonic plates clearly encountering one another as rock: "they do not melt when faced with one another, one plate does not encounter the other as a tree or as soil; the plate encounters another plate as plate, but it doesn't exhaust its being" (*Tool-Being* 222). It is because of one plate's own being as what-it-is that it experiences particular aspects of the other as what-it-is-to-it. In the same way, we can think of how our own bodies encounter artefacts and how they encounter us – our bodies are the enablers, structurers, knowers, and limiters of our experience of, and access to, the world, and, as we have seen, this deeply affects our equipment-augmented postphenomenological experience. But our artefacts similarly only encounter certain aspects of us, and only partly by design. We will shortly explore this OOO-inspired focus on the bodies of things in our discussion of what it means to know.

Finally, OOO recognises, without undermining the reality of the artefact, the role of equally real constituent components, and this gives us another insight into the resistance to the complexity of the new reading devices. When I say that the essence of an artefact, its full reality, is unknowable, it is in part because its existence is dependent across scales that don't include me. Every object has an essence because it holds together and has an effect as-what-it-is at at least one scale, but at another scale it makes more sense to think of the object as a structure, of the atoms in the process of acting out a cup, at another of quarks as acting out atoms, at another of superstrings acting out quarks, and who knows where on from there? As Harman puts it, a "hammer is always siphoned away into countless systematic unions... [T]he hammer is also made up of trillions of minuscule tool-beings [essences] which are by no means utterly dissolved in it..., a tool-being is always part and whole" (*Tool-Being* 279).[18] At every possible level, a sentient being living at that scale could encounter essence, space, eidos, time, and network: what the object is, what it seems to be, and the structures into which it is inserted and which comprise it. Bogost states this explicitly:

> Things are independent from their constituent parts while remaining dependent on them... Counterintuitively, a system and a unit [object] represent three things at once: for one, a unit is isolated and unique. For another, a unit encloses a system... For yet another, a unit becomes part of another system – often many other systems – as it jostles about. (23 & 25)

We tend to judge the boundaries of things by the causes and effects of our own level, and complex artefacts (such as watches, engines, and

---

[18] Note that this stance explicitly resists the undermining tendency to see objects as more real at these "deeper" levels. Instead, other things are *also* real, with part of their reality being their role in a larger structure.

combination locks) can confound our sense of scale, and the computer even more so because vital components function at levels beyond our comprehension without the aid of, in Ihde's terms, some hermeneutically-deployed artefact (the microscope or ammeter). Multi-component objects, when encountered as such (the gestalts that we encounter often hide their complexity), seem to be both object and network at the same time: easily pulled apart, and yet one thing – a cup is complex enough, but the computer is absurd. In comparison, the codex seems a simple mass; we hardly ever pay attention to its existence as a structure. This stems from its being able to apparently be perceived in its entirety at our scale; our bodies seem to match it all the way down to its components in a way that simply isn't the case with the internal workings of the computer, television, or e-reader. Again, this is a challenge to what it means to know, to what it means to develop expert knowledge, and it can also be seen as part of the source of unease about the "unnatural" – the materiality of printed books seems to better conform to our scale; they seem to be more, and more immediately, encounterable. But with our support from OOO we can say that this, to some extent, must be a misrecognition.

4.1.3.5.1 A faith in contact despite escape. At least to some degree, Harman maintains that "real never encounters real," that it is impossible for one object to approach another "as it is"; in short: real objects only ever encounter sensual objects. I want to emphasise how clear Harman is on this matter:

> [W]hy exaggerate and say that things cannot touch at all? Does it not seem instead that things partly make contact with each other? After all, we have been speaking all along of how humans have partial access to hammers while using them, and have also reflected on how fire touches certain qualities of cotton despite not touching the cotton as a whole. The problem is that objects cannot be touched "in part," because there is a sense in which objects have no parts. It is not as if things were made of seventy or eighty qualities and there were a mere practical limit ensuring that five or six of these qualities would always be withheld from the organs of sense. (*Quadruple* 74–75)

For Harman, because objects are always more than the sum of their qualities, simply revealing more qualities of an object cannot draw you any closer to it. I want to argue, however, against the strong form of this claim whilst maintaining its core. The core, as established above, is that one object never "satisfies another," something always retreats; a weaker form of Harman's assertion, however, would be that one thing can *approach* another, but never fully realise it. In order to make this claim I need to say more than that we can keep experiencing additional qualities of an object – Harman is very clear that this doesn't bring us closer to a thing so long as we acknowledge that objects are more than the sum of the qualities that they possess or that emanate from them. The only way that I can see such an approaching of essence being possible, therefore, is by our altering the sensual object that we do encounter in such a way that it better expresses the real object that it is always-already partly indebted to. In this way, we wouldn't somehow reveal and combine more and more qualities of the real object, but instead the qualities of our sensual object would become more closely aligned with the qualities of the ever-escaping real. In this way, I am more confident than Harman in our potential for, if not contact with real things, then an *alliance* with them, and in such confidence I must at least partly drag my take on OOO away from Harman's reading of Heidegger and back towards Husserl by saying that a real object *can, must*, be made at least somewhat available in the manner in which the sensual object manifests itself during expert use.

> Whereas for Husserl the hidden hammer-at-work might be brought into consciousness whenever we feel like it, Heidegger finds it impossible in principle to make the withdrawn reality of the hammer fully reveal its secrets. There will always be a subterranean depth to the world that never becomes present to view." (Harman, "Technology, Objects and Things in Heidegger" 3)

The truth, I suspect, lies somewhere between these positions.

I want to argue that successful action with an artefact is based upon a real alliance with something beyond mere sensual qualities; prior experience with and knowledge of an object prepares us for interactions with that object in such a way that we brush up closer to its realness. As I'll go on to argue, the very notion of expertise necessarily relies upon exactly this approach – expertise is about the minimisation of surprise through knowing; to know is to be able to act successfully. OOO's description of every object's escapological tendencies, however, also explains why the process of technologising and accumulating expertise and knowledge must always be provisional and open to corruption through surprise – my argument is certainly not that we can solve the issue of access and encounter things directly.

In *Tool-Being*, Harman, for his part, again explicitly argues against some of what I want to say:

> [t]here must be some sort of complicated way in which being announces itself in appearances; otherwise, even approximate forms of knowledge would be utterly impossible. Just how this happens remains unclear. But in negative terms, it cannot possibly be through an as-structure that would adequately mirror the things themselves, or even one that would give us a closer and closer but merely asymptotic approach to the things. The gap between the two dimensions remains absolute. (160)

Because of the absolute distinction that he sets up between the essence of the thing and the phenomenon produced during interaction, Harman sees no way for us to access any aspect of that essence through appearances (directly counter to Husserl). And I agree; we cannot get closer and closer to a real object *as a real object*. But I also believe that there is a way to embrace a variety of asymptotic approach whilst maintaining an absolute distinction between the real and the sensual, and technological use gives us the perfect example in a reading of such use that is more in line with the mainstream interpretation of Heidegger's tool-analysis. To preempt the discussion: we come closest to an object as-it-is during *repeatable successful action*. Readiness-to-hand, in the mainstream reading, is one example of such action, but, following OOO, I think that we can extend the concept out beyond the subject-object interactions of human use to the object-object interactions Harman is also concerned with. This will, in turn, allow us to talk about the ways in which technologies approach their users and even gain their own forms of expertise. I'll stick, initially, with the development of expert use in humans interacting with tools, using our familiar artefacts of hammers, printed books, and e-readers, but later on, and in Chapter 5, we can explode this out.

When we can use a hammer again and again, reliably, predictably, successfully, we must be encountering not just what the object is as-hammer, but also at least some relative of its essence which allows for the work to be done. This must be the case because the real object doesn't change even as our potentials with it do. As we saw in Chapter 3, the whole-composite (sensual) object can be moulded by experience, embodiment, discourse, and many other subtle effects, but the real object remains the same. I say the sensual object, not just its sensual qualities, because although the sensual qualities certainly change most readily (different conditions, different visual aspects, etc.), the sensual object itself also doesn't neatly persist. The computer encountered by the expert *is not* the computer encountered by the amateur; my TV becoming a teeming city, a more full black box, was in a very real way a different sensual object with different real effects. When you're a child with no knowledge of biology you do not intend the same human body as when you grow up and become a surgeon. Sensual objects are not so changeable as their qualities, but such change is what the process of technologising and of developing expertise does. Recall, for example, the invisible and then encountered "mushrooming" of the valve head in the bike engine described by Crawford in Chapter 4: two different sensual objects (a valve without flaring and a valve with flaring), not just a change in qualities anymore than a ship with sails and a ship with an engine can be the same object. What I want to argue is happening during the more informed perception is that the sensual object has become more *coordinated* or *coherent* with the real object without ever requiring an increase in our direct access to the thing itself.

Zahavi's reading of Husserl gives us a guide here:

> knowledge is not...static..., but a dynamical process that culminates when all of the profiles of the object are given intuitively...It should be emphasized too that the profiles in question do not simply refer to the appearing surface of the object, but to the givenness of all of the properties of the object, be they properties that belong to the interiority of the object or properties such as solubility that only reveal themselves when the object interacts with other objects. (*Husserl* 34–35)

For Husserl, the process of being informed about the world ceases at the moment that every profile of an object is automatically and authentically co-given from any adumbration. In this view, translated through my take on OOO, perfect knowledge would be the perfect coordination of the sensual and the real, and I agree with Harman that this is impossible. But the potential for a higher *concordance* must be true in order to explain the potential for an increase in expertise, in the possibility of repeatable successful action, with the concomitant change in the nature of the whole-composite encountered. So, against Harman, I will be arguing that the way that things appear to us, and the ways in which we encounter them more broadly, and (even more broadly) the way that all objects encounter other objects, *can* include a potentially asymptotic approach, though this needn't imply direct access. This will form a foundation for the discussion of knowledge in the next section.

It's important from the outset, however, not to mistake closeness with complexity. A more complicated description of a thing does not necessarily reveal it further as-it-is. Scientific investigation is perhaps the ultimate complexifier, moving our perception beyond the levels at which we ordinarily operate. But being aware that a hammer is made of molecules of iron and carbon, or of more fundamental atoms, or quarks, or fluctuations in superstrings, for instance, needn't draw us closer to the thing itself (at any scale) unless it also amplifies the *quality* of the interactions that we can accomplish with it. Every molecule or atom or string is, after all, also an object with a rich being of its own that recedes from contact. Complexity (i.e., the amount of qualities in, and resolution of anticipations produced by, a gestalt sensual object[19]) doesn't automatically draw us closer to the real thing itself.

The asymptotic approach to knowledge, to inadvertently knowing the real, that I want to describe is about increasing the coherence of how something appears with what it real-ly is, but it is a progression with no end and an infinite array of false paths. In a BBC interview, Vladamir Nabakov spoke about apprehending objects in a way that might be considered a folk-phenomenological account of this process, offering us a starting point to build upon:

> Reality is a very subjective affair. I can only define it as a kind of gradual accumulation of information: and as specialization. If we take a lily, for instance, or any other kind of natural object, a lily is more real to a naturalist than it is to an ordinary person. But it is still more real to a botanist. And yet another stage of reality is reached with that botanist who is a specialist in lilies. You can get nearer and nearer, so to speak, to reality: but you never get near enough because reality is an infinite succession of steps, levels of perception, false bottoms, and hence unquenchable, unattainable. You can know more and more about one thing but you can never know everything about one thing; it's hopeless. So we live surrounded by more or less ghostly objects. (qtd in Dee "Nature Writing" 27–28)

---

[19] By "amount of qualities" I mean all of the qualities that we might identify as comprising the object as it appears to us in a moment (remembering that objects are always more than the sum of their qualities and that there will always be qualities that we do not encounter). By "resolution of anticipations" I mean the vividness and accuracy of the co-intended adumbrations, that is, knowing what the television will look like as I walk around it or when I remove its casing.

I suspect that Harman would both agree with and despair at this quotation, and that's the exact position in which I want to situate the argument to come: in agreement with both the unquenchable nature of ghostly objects and the "gradual accumulation of information ... and ... specialization." This isn't exactly the approach to reality that Nabakov describes, but instead an asymptotic development of accordance between the sensual and the real that similarly underpins what it means to know.

4.2 What is knowledge?

What OOO gives us, and particularly an approach to OOO with a weakened claim for the intractability of coherence with an object's essence, is an ontological basis for a posthuman epistemology that can further our discussion of the nature of technological encounters. Below, I'll outline what a posthuman knowledge claim might look like and then put it to work, in Chapter 5, first as it is already established in a philosophical approach to evolution, and then bringing together everything that I've discussed in this book in order to show the same mechanism at work in both technological objects and technological interactions, that is, from both the user's and from the artefact's sides of use.[20]

So what is knowledge?[21] The classical claim, from Plato onwards, is an anthropocentric tripartite model: knowledge is a "justified true belief." In order to construct a posthuman epistemology we need to throw this out immediately[22] – at the very least inanimate objects (and most

---

[20] Note that I am not arguing for a division of subject and object in technological use, fully accepting that they are entangled with one another and are jointly, flatly, responsible for the effects of that use, agency not fully residing in either pole. And yet user and artefact are also real-ly different, that is, they are something real before they interact with one another and continue to be so no matter to what extent that contact domesticates either party, and even as they jointly form a new machine able to accomplish a task otherwise barred to either actant.

[21] My aim here isn't (cannot be) to write a history of epistemology. I want to talk about an approach to knowing that fits with the concerns of this book, but whether it is even a provocation to existing work on what it means to know is for the reader to decide. What the approach I will outline *does* do, however, is highlight the flattening of the hierarchy of user and artefact, draw out the implications of the claims made in this book about embodiment and relation in technological use, and offer a preliminary model for uniting ontology, epistemology, and (post)phenomenology in a way that is truly resistant to anthropocentrism (refuting human *specialness* while acknowledging human *distinctiveness*, i.e., humans are not privileged, but they are different).

[22] Not that this definition isn't already problematic in the field: in "Is Justified True Belief Knowledge?" the philosopher Edmund Gettier challenged such claims

animals) don't believe, nor do they tend to check their claims with one another. The model that I want to adopt is, instead, more pragmatic, and if not classically so then at least inspired by the American pragmatism of Peirce and Dewey (and its influence on, in particular, Ihde's postphenomenology) which sought to drag epistemology from the realm of metaphysics to the practicalities of human engagement in the world. My definition of knowledge goes even further and is therefore necessarily stripped down to the aspects of knowing which might be true for humans, animals, and inanimate objects – each of these broad genera know very differently, and possess knowledge very differently, and we'll certainly discuss some of these distinctions, but I'm more interested in outlining a fundamental mechanism of knowing, in line with OOO's approach to ontology, which applies to them all. The existing taxonomies of kinds of human knowing are therefore not thrown out, but are instead seen as subsets of a broader phenomenon.[23]

I will make three claims about knowledge, and all of them are mutually interdependent (or, rather, say the same thing from different perspectives):

- An increase in knowledge is the drawing closer of one thing's data to another thing's reality.
- An increase in knowledge is an increase in data that enables repeatable successful action.
- An increase in knowledge reduces the flow of information between one thing and another.

Let's take each in turn.

---

as comprising what it means to "know" in any satisfactory way. In that paper he offered a series of what have become known as "Gettier problems." An example: a man, Smith, applies for a job. He has a justified belief that Jones will get the job and that Jones has ten coins in his pocket (he may have overheard that Jones is the favourite for the job, and he may have had an opportunity to count the coins in his pocket). This leads him to the "knowledge" that "a man with 10 coins in his pocket will get the job." As it turns out, Smith gets the job, but it also turns out that, by coincidence, he also had ten coins in his pocket. His claim based on justified belief turns out to be true and yet this seems an unsatisfactory description of knowledge. What I will go on to argue resolves such Gettier issues, partially by relying on a condition of the *repeatability* of success – with repetition, the probability of successful action based on poor or incoherent data significantly decreases; felicitous conditions rarely remain consistent for long.

[23] Again, it must largely be for others to decide, if they accept the claims here, the compatibility with prior epistemological work.

4.2.1 Knowing is getting closer

The first claim is a result of combining Harman's OOO with my discussion of expertise which satisfies the human aspect of increasing knowledge. The expert knows more about technological equipment because the sensual object that they encounter coordinates more closely with the real object – expertise and knowing, to some extent, are defined by the ability to produce increasingly "accurate" sensual objects with the data for this accuracy provided from encounter or from storage. The expert craftsman using a hammer to drive a nail, who is rarely surprised by the encounter, must necessarily be producing sensual gestalt objects of hammer, nail, wood, and her body combined with the hammer (as one thing in Incorporation) that cohere with the realities of those things. If the sensual objects had too little in common with the real objects that they are mistaken for, then action could not take place. An amateur, who is constantly surprised by how the hammer and wood and nail and (maybe particularly) her body react, produces and acts on sensual objects that cohere far less with reality – this is what characterises amateur experience.

Andy Goldsworthy, a sculptor who works almost exclusively with materials found in the landscape in which he works, offers sympathetic folk-phenomenological descriptions, in this regard, of the development of his expertise with and knowledge of his sources. Firstly, he acknowledges the parity between his art practice and craft expertise more generally: "I understand snow and leaves and feathers and mud and sticks and stones a little bit like the way a carpenter will understand wood, because he's worked with it" (166). He also recognises that his interactions lead him towards a greater knowledge of a thing, but that this knowledge is always hard won and insufficient: "All my work concentrates on a particular aspect of material or place. The grass stalk is hard, brittle, hollow and fractures at angles; the seed-head is supple, thin, strong, whippy. It takes many works to come to some understanding of 'stalk,' let alone 'grass'; it will take many more" (162). Finally, there is some recognition that he is revealing something real through practice, something that transmutes initial perception: "The most rewarding thing ever said to me was by a Dutch woman of a shape I had carved in sand. She said 'Thank you for showing me that was there.' That is what my work does for me myself, the discovering 'what was there.' If it does for others, then so much the richer" (163). I don't believe that Goldsworthy ever thinks that he has solved the mystery of a thing, but what is good and important about his practice for him, and for those who spend time with his work, is that sense of getting somewhere closer to reality without a solution in sight. In the terms that I want to set out here, his increasingly expert understanding of the materials is tied to an increase in the coherence of his prehensions.

I'll define more precisely what I mean by "data" in the next section, but for the moment we can simply say that experts must store data about, for example, hammers and the act of hammering within themselves, and that this is evidenced by their being able to produce more accurate and reliable sensual gestalt objects during technological encounters than an apprentice – something must endure over repeated engagements. This kind of data, data that can be acted upon because it more or less closely coheres with reality, is what I want to call "knowledge".[24]

An important clarification must be made, however, in light of the insights from both radical embodied cognition and postphenomenology: human knowledge-data isn't just stored in the head. This continuation of the radical approach can open the doors to a more posthuman attitude towards knowing – the expert's knowledge of, for instance, hammering is partially formed of data inside her brain, but relevant data is also in the hand and arm that operate the tool, in the tool itself and its feedback, and in the act of hammering with *this* hammer in *this* moment in *this* milieu. The expert, in a very real way, changes her knowledge of the hammer that she holds and of the act of hammering in the moment of using the equipment; experience always prompts new data to draw on, and this tends to strengthen the coherence of sensual gestalt and real object. Such data is always dispersed across the contextualised 4EDS soft-assemblage – it is the *technological encounter* that knows best, not just an expert human brain.

The cognitive archaeologist Lambros Malafouris, to different ends, describes such an assemblage with regards to pottery thrown on a wheel:

> We should assume...that every mental recourse needed to grow a vessel out of clay may well be extended and distributed across the neurons of the potter's brain, the muscles of the potter's sense organs, the affordances of the wheel, the material properties of the clay, the morphological and typological prototypes of existing vessels, and the

---

[24] It is not, however, that all data about the world is knowledge – to pre-empt my third claim: you can store data about the world that doesn't enable your action (e.g., because it is actively false). Knowledge, as I'm describing it, is only what provides the capacity to act, and, as we will see, this means that it is the kind of data about the world that can be tested.

general social context in which the activity occurs... I do not mean to deny that an intricate computational problem may well arise for the brain the moment the potter touches or is touched by the clay; I simply mean to emphasize that part of the problem's solution is offered by the clay itself, without any need for mental representation. (213 & 219)

Malafouris reveals his indebtedness to Chemero here and gives us a great model for knowing. How does the expert potter know how to bring about a pot? In part there is something in her head: when the clay and wheel are simply contemplated (mainstream reading of presentness-at-hand) knowledge-data is predominantly brain-bound with additional input from the body, situation, and context. But part (maybe most) of knowing how to make a pot is also in the act of doing: during the act of expertly throwing the clay (mainstream reading of readiness-to-hand), knowledge-data, which is drawn upon in order to produce the sensual object encountered, is further distributed across brain, body, tool, situation, and context, and this activity (again, against Harman's claims) tends to be richer and of a higher quality than mere contemplation (at least during successful use).[25]

Expert activity, primed by a wealth of prior experience and receptive to real-time data inputted from the world into the system of which the expert is a part, produces the most coherent sensual objects that humans can accomplish. Malafouris again: "there is a great deal of approximation, anticipation, guessing, and thus ambiguity about how [a] material will behave. Sometimes the material collaborates; sometimes it resists. In time, out of this evolving tension comes precision and thus skilfulness" (176). This "evolving tension" I have called "technologising" and will go on to call "informing" – we can now see that knowledge and expertise, at least as I'm describing them, are much the same thing, and technologising is a distinctive kind of accumulation of knowledge and expertise in tandem with equipment, an encounter that has particular describable features (as outlined in Chapter 2).

I realise that this view of knowledge is already profoundly unintuitive in some aspects, so for clarity I want to make a final point on this radical and postphenomenological spread of knowledge-data that, whilst still about the use of supporting equipment, is more in line with what we might see as a classically epistemological issue: failing knowledge.

---

[25] This also highlights the importance of the testability of knowledge that we'll return to shortly.

Naomi Klein, in *The Shock Doctrine*, details the tragic events that have shaped much of Gail Kastner's life. Kastner was subjected to intense psychological "treatment" by Dr. Ewen Cameron, a psychologist working on mind control and "reprogramming" for the CIA during the Cold War. Cameron's techniques – sensory deprivation, isolation, massive administration of electroshocks, and cocktails of drugs – resulted in Kastner experiencing bouts of psychosis, massive depression, and a loss of memory that manifests itself in amnesiac symptoms and the impossibility of forming concrete new memories. One of Kastner's coping mechanisms, in the face of the effects of her abuse, is particularly telling for our current discussion. The following is from Klein's first meeting with Kastner:

> [S]he goes into a minor panic when I ask her for four inches of space for the recorder. The end table beside her chair is out of question: it is home to about twenty empty boxes of cigarettes, Matinee Regular, stacked in a perfect pyramid... It looks as if Gail has colored the insides of the boxes black, but looking closer, I realize it is actually extremely dense, miniscule handwriting, names, numbers, thousands of words... Over the course of the day... Gail often leans over to write something on a scrap of paper or a cigarette box – "a note to myself," she explains, "or I will never remember." The thickets of paper and cigarette boxes are, for Gail, something more than an unconventional filing system. They are her memory. (27)

My interest in this description, beyond the vital story of abuse that Klein tells, lies in Kastner's attempts to build herself a new memory. She doesn't write down mere scraps, or select essentials, she attempts to capture a lot of data, some of which we can assume is trivial from the sheer amount being recorded. As our memories, when they function as we expect, are far from picky, so Kastner's memory is expansive. As we layer our thoughts, often to the detriment of clarity, so does Kastner. There is no sense, at least in the narrative that Klein presents, that Kastner initially set out to reconstruct a memory outside of her head, instead she simply started to use lists in order to account for a failure in her more bounded thinking machine. But now she relies on knowledge-data spread far out into the world.[26]

---

[26] For any reader familiar with Clark and Chalmers's discussion of "The Extended Mind" this story must resonate with their example of Otto and his notebook. Clark and Chalmers are interested in the extension of mind (or, if we

When someone is in Kastner's position, or if they have Alzheimer's, or suffer from some other degenerative neural condition, then the knowledge-data held within the brain necessarily diminishes with the brain's deterioration. The coping mechanisms that are developed in the wake of this loss are, I would argue, not about changing what it is to know, but, rather, *where* it is to know. Routine, labels, lists, human and animal helpers, they all act as distributed aspects of the knowing *system* by migrating data to the world, to the body, or (via plastic redistribution) to different areas of the brain.[27]

Whilst we might talk about individual humans knowing things, this is always at least partially, and often profoundly, in concert with the contextualised 4EDS architecture that we have already established – the system knows. In this way the locus of knowledge is already shifted from being neatly anthropocentric.

---

prefer, cognition), but my discussion of knowledge here is designed not to be reliant on cognising subjects being present in a system that possesses knowledge. The Extended claim with regard to Otto can therefore be seen as a subset of what I'm discussing here:

> Otto suffers from Alzheimer's disease, and like many Alzheimer's patients, he relies on information in the environment to help structure his life. Otto carries a notebook around with him everywhere he goes. When he learns new information, he writes it down. When he needs some old information, he looks it up. For Otto, his notebook plays the role usually played by a biological memory. Today, Otto hears about the exhibition at the Museum of Modern Art, and decides to go see it. He consults the notebook, which says that the museum is on 53rd Street, so he walks to 53rd Street and goes into the museum. Clearly, Otto walked to 53rd Street because he wanted to go to the museum and he believed the museum was on 53rd Street... [I]t seems reasonable to say that Otto believed the museum was on 53rd Street even before consulting his notebook. For in relevant respects the cases are entirely analogous: the notebook plays for Otto the same role that memory plays for [anyone not reliant on a notebook]...; it just happens that this information lies beyond the skin (12–13).

[27] Further support for this claim can be found in Harris et al., "Couples As Socially Distributed Cognitive Systems." As Alex Fradera notes in his review of this paper,

> older adults tend to experience the greatest memory difficulties with first-hand autobiographical information, rather than abstracted facts. This is exactly where...couples gained the biggest benefit from remembering together, as evidenced by performance on [an] in-depth event recall task and...spontaneously emerging anecdotes. It's possible that as we grow older, we offset the unreliability of our own episodic systems by drawing on the memorial support offered by a trusted partner. This might explain why when one member of an older couple experiences a drop in cognitive function, the other soon follows. Our memory systems are more of a shared resource than we realise (Fradera).

4.2.2 Knowledge enables repeatable successful action

If my first claim is that an increase in knowledge is a drawing closer of one set of data (however it is stored) to the essence of a thing, then the second describes what can be done with that data, that is, we can say that to possess knowledge is to possess data that enables repeatable successful action. This understanding of knowledge has a strong connection with American Pragmatism and particularly Dewey, James, and Peirce's understanding of knowledge as being *for* something. As Ihde describes the pragmatist concern (also influential for Chemero), the "emphasis is on *practice*, not *representation*" (*Postphenomenology* 9, emphasis in original). Similarly, when I suggest that there is a coherence between knowledge-data and the world, I'm not suggesting that there is a neat picture (representation) of the world in the brain, instead I'm saying that there can be data held in systems which allow those systems to act successfully (where success has an evolutionary flavour in being able to occur again under the same or similar conditions).

As with my definition of technology in Chapter 2, this stripped down approach to knowledge is inspired by Pitt, who also draws on pragmatism: "I will opt for an operational, call it a pragmatic, approach to knowledge. This means that I will look for the hallmark of knowledge to be successful action. If, on the belief that x causes y, when I do x, y happens consistently, that is good enough for me" (xii). And later:

> [T]he ultimate test of what is to count as knowledge is determined by our ability to act successfully on that knowledge.[28] If the world is reported to be a certain way, the final test of that candidate claim will be the success of an individual or group acting as if the world were in fact that way. Not only is knowledge determined by the limits of action in this fashion; its purpose is action. Thus we seek to discover or uncover the way the world is in order to make our way around in it better. (5)

In the human terms of traditional pragmatism that Pitt draws upon, to know is to enable action in the world. For our purposes here, however, we need to take this idea further in order to extend it out to a posthuman understanding of knowledge, but also to understand that knowledge doesn't just allow action, but particular *kinds* and *qualities* of action. If I "know" that a stretched sheet can safely break my fall from a burning

---

[28] Pitt cites Peirce's "The Fixation of Belief" in this regard.

building, then that enables me to act in the world: I jump out of my window with the knowledge that I'll be ok. When it transpires that the sheet can do no such thing, however, then it becomes clear that I didn't know anything at all; I simply possessed data and acted on it. Knowledge is about getting closer to things, and it manifests itself by allowing actions based upon the same data to occur again and again (under the same conditions[29]) with the same or highly similar results.[30] By a "kind" of action I mean simply that knowledge-data facilitates *relevant* action, for example, knowing that a spider is venomous has no bearing on whether or not I can drive a car. By "quality" of action I mean the predictable repeatability of the same results on the same data; it is the quality of actions that is most significant for our understanding of expertise. For these reasons, knowledge, as I'm describing it, isn't binary: it's analogue, granular. You never simply know or don't know; instead you only know so much. To ask whether you know the meaning of a sentence isn't a coin toss, it always takes into account the subtlety of interpretation and explicitness of mistake – in the same way, whilst some knowledge feels so certain that it appears binary, this is never the case.

If I know how to hammer a nail then I can hammer nails again and again and again. This is a fairly neatly intuitive sense of knowing, despite the complications of (a) acknowledging that knowledge is always provisional, I can't know everything about anything, and (b) what knows, in the end, isn't "me," but a cognising system that includes me. My knowledge of how to hammer a nail is tested in action, with constant small refinements occurring in response to *this* nail, *this* hammer, *this* piece of wood, *these* conditions of humidity,

---

[29] Knowledge is based on the necessary faith that conditions can be repeatable to some level of accuracy (scientific investigation is clearly predicated on this); if every interaction was entirely novel, we could not act, and in a very real sense we would know nothing.

[30] In this regard also see Davis Baird's work in *Thing Knowledge*:

a "true wheel" is not true simply because it properly conforms to a particular form; a true wheel spins properly, dependably, reliably. A wheel that is out of true wobbles and is not dependable. Ultimately it will fail. This sense of 'truth' picks out those contrived constellations of materials that we can depend on. A public, regular, reliable phenomenon over which we have material mastery bears a kind of 'working knowledge' of the world and 'runs true' in this material sense of truth. (122)

Presenting ideas such as this, Baird has been an influence on my thinking about how knowledge can be stored in things.

this level of exhaustion in my muscles from prior work, etc. I say that "I know how to hammer a nail" because prior experiences of hammering have given me the ability to produce high quality whole-composite sensual objects of hammer, nail, and wood that I am able to base actions upon. The more expert that I am, the greater the coherence between the sensual objects that I produce and their real escapological depths. But the act of hammering is in concert with wood of unknowable densities; nails of unknowable strength; atmospheric conditions of unknowable effect – with every strike I reassess, adjust, and I alter some of the sensual objects that I intend (adding the quality of density to a previously assumed softer wood for instance) in order to try and facilitate coherence, to enable repeatable successful action. Human knowledge, in this conception, is at best good enough in a moment, an asymptotic edging-towards that is reflected by the success and repeatability of an activity. A coherent sensual object, therefore, is a conception where what something is-to-us matches, at action-facilitating points, aspects of the real thing intended. If something changes too quickly, or if we never act with it on multiple occasions, then we can never tell whether we have brushed close to the thing in itself. But success is telling.

David Hume's scepticism of our capacity for inductive reasoning, however, is resonant with the importance I want to place on surprise revealing the dearth of access that we have to a thing: just because something has been successful once, twice, or a hundred thousand times simply cannot mean that we know it will be true tomorrow; it merely suggests that it is probable, and maybe not even that (though Hume finally had a faith in our instinct for what will endure, another form of expertise). As such, when I say that repeatable successful action implies some kind of access to an object it can only be in the weakest sense possible: that the object allows this kind of interaction, and this only holds up to the point that it doesn't and we're surprised. We can speak of a total or complete knowledge only when we can say: "if these exact conditions occurred again then the same outcome would arise during this exact activity." Knowledge isn't in the description or representation of the conditions, but instead in the repeatable potential for actively matching data, however and wherever it is stored, with a real state in the world. Again, we see the insufficiency built into this conception of knowledge – perfect knowledge is always dependent on the exact replication of conditions and the material issue of ideally storing some form of corollary about those conditions in some system (data are always metaphorical).

4.2.2.1 Data

Before we continue, it is worth clarifying what I mean here by "data." Again, this must be a simplified approach that fits with my fundamental posthuman application. We could be more specific about types of data in particular situations (sense data, database data, DNA data, etc.); as I am discussing it, however, data is any kind of *difference*. If something is red, then it's not black or white or blue; if something is here, then it's not there; if something is six feet tall, then it is no other height. These are all data points, facets of distinction about something in the world. In OOO terms we could even call them "qualities." Data, in this sense, is therefore always bound up in real objects in the world (and this view is compatible with OOO in that there are an infinite variety of kinds of difference, the majority of which we can never know).

But more or less correlating data must also be stored in the things that do the encountering, like humans, in order to produce sensual objects with varying degrees of coherence to reality. To initially stick with the simpler anthropocentric sense, when I know how to use a hammer then there must be some kind of storage of data in my head that correlates with data in the world (or I at least I believe that it does; as ever it gets tested by action). Note that this needn't be a mental representation of the world, simply that there is something in my head, some set of qualities, that I can act upon. My aim here is not to discuss the processing or storage of data, with regards to humans and animals those are questions for Neuroscience, but that the processing and storage of data must occur is evidenced by the potential for action.

So there are data (qualities) in things in the world. As the philosopher of information Luciano Floridi describes this kind of data, it is a lack

> of uniformity in the real world. There is no specific name for such "data in the wild." One may refer to them as *dedomena*, that is, "data" in Greek...Dedomena...are pure data, that is, data before they are interpreted or subject to cognitive processing. They are not experienced directly, but their presence is empirically inferred from, and required by, experience. (*Information* 23)

Floridi, with no hint of alliance with OOO, gives us a perfect description of qualities here – dedomena are the data that must be out there, the differences that make all experiences, and all objects, what they are before interpretation.

In addition to such dedomena "in the wild," there are the data in the encountering things (as described above) – most simply there is data in my head that enables me to say: "I know." I would argue that such data in the head (or knowing system) is also best thought of as dedomena as Floridi describes it – data in the human that does the encountering is similarly a lack of uniformity, one system of brain states over any other, but it is a change that happens to have been produced in response to data in the world. Let's call the data in what is encountered in the world "dedomena" and the data in what does the encountering "data," but only for clarity – *they are both the same thing*. The real hammer has dedomena (qualities); my brain has data about the hammer (qualities). This second batch of data is part of how I produce gestalt sensual objects – we are speaking about the coherence between two datasets. Stored knowledge-data is special in that that it is data that can reliably be put to use, that is, an increase in knowledge-data is an increase in the coherence of stored data with some other system of dedomena.

For this reason, stored data is always data regardless of its coherence with dedomena. Data itself isn't any measure of coherence; data is just a set of qualities of an object or system of objects (e.g., a hammer, a brain, or the hammering ensemble).

4.2.2.2 Posthuman repeatable successful action

To expand the second assertion about knowledge out to a posthuman approach we need to ask: "how can objects act repeatably and successfully?" For Harman, a cup placed on a table encounters aspects of the table without satisfying the table's full reality; the cup reduces the table down to a set of encountered sensual qualities. With the description of data above, and of knowing as repeatable successful action, we can rephrase this as "a dataset within the cup coordinates with some specific set of dedomena of the table" – there is a concordance between the data of each object demonstrated by repeatable successful action (and this is what affordances are based upon). Whilst the cup may not satisfy the rich reality of the table (any more than a human could in their own encounters), the data stored within its own real body at once produces the sensual gestalt of the table that it encounters and must, in turn, coordinate at action-facilitating points with the real table – my contention is that this is the same in kind, but not in specificity, as human knowing. Humans may be active in striving to take inputs from the world and store relevant data for future action, but the end result is the same, tending towards the concordance of two real objects without their ever encountering one another as real, as what-they-are.

To reiterate, this isn't to suggest that cups know like humans know, but that there is a parity in that they both possess qualities that, in their testability, can be described as "knowledge." Making a similar point, and drawing on work from Whitehead, Harman argues that

> all human and non-human entities have equal status insofar as they all prehend other things, relating to them in one way or another... This does not entail a projection of human properties onto the non-human world, but rather the reverse, what it says is that the crude prehensions made by the minerals and dirt are no less relations than are the sophisticated mental activity of humans. Instead of placing souls into sand and stones, we find something sandy or stony in the human soul. (*Quadruple* 46)

What is unique about human knowledge is not that it exists, but that we have the ability to store and draw on a great diversity of more or less coherent data about the world in dynamic systems that we can actively and rapidly shape. This isn't true for the cup, and yet the quality of its coherency with the world (for particular data) is greater than anything within our brains. Our bodies, of course, encounter the table with as great a coherency as the cup; they are equally objects, but our complex cognition will always include a raft of interference as we go about producing distinct sensual objects.

4.2.3 Knowledge stops the flow of information

My third claim for a posthuman approach to knowledge is that knowing depends on a drop in the flow of information between the knower and the known. As with our brief discussion of epistemology, it can't be my task here to outline a complete take on information theory, so I'll again limit my discussion to an underlying mechanism for information, one which partly draws upon some of the themes of Claude Shannon's work in communication theory from the mid-twentieth century.[31] My aim is to complete a language of knowledge, data, and information that I can go on to use in Chapter 5 to discuss how our artefacts are informed by

---

[31] This view of information has at least some of its roots in Shannon's hugely influential "The Mathematical Theory of Communication." Shannon sought a similarly stripped-down view of information, one that ignored the meaning of a given message in favour of quantifying the amount of uncertainty, or (potential for) surprise, that its every element removed. For more on Shannon see Floridi's *The Philosophy of Information* or Gleick's *The Information*, both of which continually return to the implications of Shannon's ideas.

the world that they find themselves in and how they come to know that world better.

Intuition actually gets us much of the way to the sense of information that I want to outline – were we to perfectly know something, if our sensual gestalt objects were wholly coherent with the world, then we would not be required to store any more data, or to effect any change in the data that we possessed. If we define the process of "informing" as "the acquisition of meaningful data" (and "information" as the data acquired, that is, information is always "about" or "because of" something), then total knowledge can be seen as the cessation of this process. When action occurs (or could occur), but the flow of information ceases, then we wholly know.

Whenever we act, when we hammer, or throw pots on a wheel, or read from an e-reader, we learn a little bit more, about ourselves, about the materials that we work with, about the actions undertaken. We come to know more; as I've described it, we produce more coherent sensual objects that allow for repeatable successful action in the future. "Informing" describes a process by which anything, human or artefact in our case, can come to know more – it is the provocation of the storage of data in any system in response to any interaction. That data "about something" is "information" and it most often forms the basis for knowledge, data "for something."

When I use a hammer I am constantly reacting to all of the unexpected effects caused by the real objects exceeding the sensual objects that I intend – the hammer behaves like *this* and the wood like *that*, the nail like *this* and my body like *that*. This experience is information rich, I am constantly being informed about these things and therefore constantly modify my stored information-data, and this causes me to develop more skill over time (knowledge-data for). An expert craftsman is less surprised by the things that she encounters and so less information flows from the world, that is, the information-data that the craftsman holds (or, rather, that the system of which she is a part holds) remains fairly constant while allowing successful action (i.e., it is also knowledge-data). The cup sitting on the floor, from our posthuman example of knowledge above, changes very little in reaction to the surface that holds it; its limited, but high-quality coherency prevents the flow of information – it is, a priori, informed about flat surfaces because of what it is, and that information-data also functions as knowledge-data for its successful action. If the surface dramatically heats up, however, and the cup starts to melt or smoulder or char, then the data that it holds alters – it is informed by the world and acquires new information-data. But remember that this data only becomes knowledge if it allows future repeatable successful action. It is in this way that information is not the same as a growth in knowledge; it is the provocation of data storage, not necessarily of actionable data. Similarly, our expert may learn a bad technique, thinking that it is more effective, but actually limiting her potential for repeatable successful action over time – she has been informed, but she has not improved her knowledge (we may still want to describe this as "misinformation"). It is of the same order of effects, but not identical in kind – new data storage has been provoked, but it is only if it's actable that I want to call it "knowledge."

4.2.3.1 Posthuman information

To take a posthuman approach to this view of information as the reduction of surprise, we can turn to what Floridi describes as "environmental information." For Floridi, environmental information is the linking of a system of dedomena, that is, of non-semantic information in the world, with some other system "in such a way that the fact that *a* had a particular feature *F* is correlated to the fact that *b* has a particular feature *G*, so that this connection between the two features tells the observer that *b* is *G*" (*Information* 33). Floridi offers the example of litmus paper (*a*) and a solution to be tested (*b*) where the paper's turning red (*a* has feature *F*) informs an observer that the solution is an acid (*b* has feature *G*) (33). There is an interesting fixation in Floridi's account on the observer who cannot access the world, but can brush close to it – the litmus paper, an example of a hermeneutic mediation in Ihde's terms, offers a nice metaphor for the production of a sensual gestalt object.

But I'm more interested, for the moment, in the way that the acid of solution *b* imparts a storage of data in litmus paper *a*, in how we can say that *b* informs *a* regardless of the presence of a human observer. An observer can draw on that information-data as knowledge-data for action, that is, the change in colour can inform us, cause us to store more data that we can act upon. But the litmus paper also possesses information independent of human concern. Dunked into the same solution again there will be no further change, no new data stored, and there is a correlation that remains of a high quality; that the paper indicates that the solution is acidic even though it cannot alter any further doesn't negate the information stored even as the limit for storing new data along this axis is revealed. Dunked into an alkali solution (*c*) after acid *b* and the paper still doesn't change because its capacity to take on new information is inhibited by its chemical structure – it is not informed and offers no actable data about the pH of the solution that it's currently in; it knows something else entirely.

In this way, and in line with OOO, the acid's real data is always unavailable to the litmus, but it can still prompt the storage of data in *a* by the process that I want to call "information." *a* has a "knowledge" of *b* to the extent that its stored data coheres in some fashion with that which is stored in *b*, and we can clearly see that this will never be perfect or total. Litmus doesn't "satisfy" acid, it reduces it to a narrow set of sensual aspects and encodes data (once) based on that reduction. To the extent that it may provide repeatable successful action it is knowledge, but litmus tells us nothing about what happens when we dip our finger in the jar; that is knowledge that we have to get for ourselves.

4.3 Coda

In this chapter I've tried to demonstrate how an object-oriented approach to expertise and knowledge can offer us a way-in to further understanding what it is that we encounter during a technological interaction and how that gestalt sensual object can be shaped. I've also laid the groundwork for a posthuman approach to technological artefacts' knowing in themselves, how their bodies can be seen as holders, deployers, and receivers of data, how they can know and how they can be informed. The aim has been to move towards a theory of knowledge and information that is sensitive to the complexities and distinctiveness of human cognition, but also recognising it as a subset of a broader phenomenon of imperfect contact and slowly asymptotic awareness. It describes a process by which humans and technologies can shape one another without explicitly trying to, a co-evolution, and it describes a system of expertise that is more than just brain and body. It also demonstrates the way that the intimate human technical relationship means that discussions of technology take us into our most fundamental questions about being, knowing, and acting.

In the final chapter we will focus on these posthuman processes of taking on information and knowing, putting these ideas to work by looking at how artefacts, and particularly those best suited to technological interactions, come to know their environment of use, and how this can be seen as an evolutionary process.

5
What Everything Knows: Technologies as an Embodiment of Knowledge

In the last chapter we discussed a stripped down approach to knowledge, information, and data that was compatible with OOO's essential insight into our fleeting brushes with reality as we become more familiar with things in the world. This final chapter brings together what we've considered so far, from the definition of a more "natural" conception of technology to the nature of the objects that we encounter, but it focuses on the kinds of knowledge that are stored in our artefacts, particularly those best suited to being at the heart of technological interactions because the data that they possess, the things that they know, coordinate well with their human users. We'll begin by looking at how the development of technologies can be seen as an evolutionary process and then twin this with our discussion of knowledge by focussing on how such evolutionary processes imbue artefacts with particular data in relation to their environment.[1] Finally, I will argue that every technological interaction can be seen as an encounter between two knowing components that are both flatly responsible for the successful action and inform one another, generating new expertise, without ever really meeting each other as real things – the human and the tool, despite the

---

[1] I won't be dealing with the invention of artefacts here; a great discussion of how artefacts come to be by drawing on what has come before can be found in W. Brian Arthur's *The Nature of Technology: What It Is and How It Evolves*. Instead, I am predominantly focussing on how already invented technologies evolve during their repeated use. The methods of invention are hugely various, from sudden inspiration to graft and slow mutation, but the mechanism of development once the artefact has started being put into practice is, I want to argue, much more stable.

seeming intimacy of Incorporation, escape from contact; they remain, to extend Nabakov's idea, more or less ghostly to one another.

5.1 Evolution of technology

In order to develop the last chapter's views of knowledge into a truly posthuman approach, one which both draws on and satisfies the assertions of OOO that we've discussed so far, an approach that enables us to talk about the importance of knowledge for technological interactions not just in terms of human expertise, but also of the object's knowledge of the user, we need to take a brief detour. I want to take seriously the idea that artefacts are embodied and are affected by their embodiment in a way similar to humans and other biological forms, and part of this seriousness is to make the argument that artefacts are the products of evolutionary forces. If such an assertion is persuasive, then it opens technology studies up to a branch of Biology-influenced Philosophy, evolutionary epistemology, which deals with the knowledge (not just the data) coded into the embodiment of biological things. But how can such a claim be justified?

5.1.1 A fundamental mechanism

As with our prior discussions I want to focus on a stripped-down mechanism that can be seen as underlying a wide variety of cases. I'll therefore outline a basic Darwinian approach that, while it can certainly be nuanced in its specifics, forms an accurate core across instances.[2]

Darwinian evolution begins with the meeting of two things: an individual organism (let's use an animal for our example) and an environment. The environment is the surroundings that the animal spends its time in, including everything within that space: members of its own species, plant life, weather, prey, predators, landscape, etc. And this animal is the expression of the genes passed on to it by its parents. Genes can be thought of as the fundamental but malleable instructions for building a body; they give a developmental range for every aspect of the animal. Genes are like a line in a recipe that recommends "a good pinch of salt" – the recipe guarantees that salt will be in the final dish, but the flavour will differ slightly or dramatically according to the interpretation of the instruction. In much the same way, the expression of

---

[2] For a more comprehensive discussion of evolution in Biology see Douglas Futuyma's *Evolution*. See footnote 36 in Chapter 2 of this book for more on Darwinism as a universal mechanism.

genes can vary across a range for each new offspring, and "epigenesis" is the term used to describe the individual's expression along such ranges. For instance, a set of genes could state that a developing claw will grow between one and three inches long, and between half an inch to an inch wide; there is no specific gene for a two inch by one inch claw for example, genes just set boundaries for development with "additional information...gained from the environment during epigenesis – genes give some liberty for development" (Vehkavaara 213).

Environmental pressures, from what the mother eats during pregnancy to how much sunlight the young animal gets, can affect the way that genes are expressed, causing different selections to be made across the diversity of ranges made possible by the animal's genetic instructions. The sum of genes that an animal carries within it, inherited from its parents, are called its "genotype," whereas the epigenetic selections from the ranges that cumulatively make up the specific developed animal are called the "phenotype" – "the phenotype is the expression of...[genetic] information in the flesh-and-blood individual that develops via a series of highly complex interactions with the environment" (Plotkin, *Darwin Machines* 95). In evolutionary terms, the phenotype is "successful" if it survives long enough to go on to reproduce, to pass on a share of its genes by producing young after resisting the threats and utilising the affordances of its environment. This, then, sets the stage for evolutionary effects. Reproduction and the influence of the environment on development are not the evolutionary process; Darwinian evolution instead occurs because organisms and environments aren't fixed.

Environments can change in many ways, but let's take the example of a shift from a moist to a dry climate. In a moisture rich landscape many plants will have the perfect conditions to thrive, and our animal, a four legged herbivore mammal, has plenty to eat low to the ground. As such, the fact that it is a relatively small creature doesn't stop it from being able sustain itself for long enough to reproduce. It is born with a complex blend of its parents' genes, and those genes are further expressed through its development (*in* and *ex utero*) within an environment which affords sustenance. If, however, the climate were to dry out, then the grasses and plants that our animal eats would start to disappear and it would find itself struggling to acquire food. If it starves too quickly, then it won't be able to reproduce and its particular combination of genes will disappear. Unfortunately, with all of the grasses dying out in the newly arid climate, isolated trees instead start to thrive and take over with their increased access to the remaining available water. Over time, the average height of the available vegetation migrates upward, and our animal's shortness has a real impact on its survivability.[3]

Reproduction is similarly variable. As stated above, a process of epigenetic selection from a myriad of genetic ranges results in the phenotypic expression of the animal. If our animal's parents had produced a perfect clone of themselves (or a simple blend of their pairing), that is, if they passed on exactly the same genetic material that they carried, then their genetic ranges (if not their expression) would stay the same. But this isn't what occurs: in the same way that the environment is unstable and continually changing, so reproduction is not the perfect transmission of a genotype. Instead, mutations occur, changes in genetic information that alter the ranges a gene (or set of genes) will dictate. In a world where vegetable matter is growing further and further from the ground those animals that can reach it will survive longer, be stronger, and have a better chance of reproducing. In biological terms they are more likely to be successful, they are "fitter," and we can think of this as their being a better fit to the environment that they find themselves in – they concord. The affordances of the world and their embodiment match in some way, and this will become significant (and hopefully already begins to resonate with the discussion of information and knowledge in Chapter 4).

The range of possible expressions which produce, for instance, our animal's neck are dictated by various sets of genes. In the drying-out world the animals which, for whatever reason, express neck length at the higher reaches of these ranges will survive better, and the species' overall neck length range will therefore tend towards genes which more often express longer necks. Once in a while a mutation will occur which extends the range of neck lengths beyond its previous limits. If an animal carries these mutated genes, and during epigenetic development expresses towards the new upper limit of neck length and therefore thrives, then these new rogue genes will become a part of the gene pool; their continued expression and success represents a change in the species.

For our purposes here we can therefore see Darwinian evolution as divided into three phases enacted upon collections of individuals

---

[3] I have no doubt that the actual reason for vegetation increasing in average height is often very different to this, and certainly more complex; my point is simply that environments change as a response to the cumulative pressures on the organisms within it, and this, in turn, places different pressures on those same organisms. In the language of the last chapter, environments are systems of information-data about enacted pressures. Here, however, we will focus on the information- and knowledge-data bound up in individual organisms.

in environments: *variation* (due to mutation), *selection* (due to pressures from the environment), and *reproduction* (due to an individual's success).[4]

5.1.2 Extending the mechanism

With an understanding of what I intend by "evolution" in place we can now consider whether material equipment, artefacts, can be thought of as abiding by the same principles, using printed books as our example. I should note from the outset that such an idea is not novel in and of itself, the biologist Peter Medawar going as far as to say that "[e]veryone has observed with more or less wonderment that the tools and instruments devised by human beings undergo an evolution themselves that is strangely analogous to ordinary evolution, almost as if these artefacts propagated themselves as animals do" ("Technology and Evolution"). But Medawar's ensuing example marks a distinction between the evolution of artefacts that I would like to suggest and that as it is more frequently discussed: "Aircraft began as birdlike objects but evolved into fishlike objects for much the same fluid-dynamic reasons as those which caused fish to evolve into fishlike objects." To me it seems that there is a mistake of environment here, of how and why artefacts evolve. Fish evolved due to the pressures exerted upon them by an aquatic existence, but "species" of aircraft feel no such pressures from the air itself; the three stages of variation, selection, and reproduction have no identified corollaries in this example, and this distinction is one that I think needs to be addressed if we are to use that term "evolution" with any validity.

The concept of the evolution of human culture has its supporters and detractors who typically question the ideas of cultural and material evolution as part of the discussion of the evolution of ideas,[5] or of knowledge gains in science specifically, rather than the evolving physical bodies of artefacts.[6] Detractors of the idea tend to appeal to a gross

---

[4] To further flesh out this understanding, two good introductions to these stages of evolution and their history as concepts are Mark Ridley's *Evolution* and Carl Zimmer's *Evolution: The Triumph of an Idea*.

[5] William James' 1880 lecture "Great Men, Great Thoughts, and the Environment" is almost certainly (in no small part due to the proximity of its publication to Darwin's own work) the first theoretical application of evolutionary mechanisms to the progress of thought and ideas.

[6] For more on this discussion see, for instance, Walter Vincenti *What Engineers Know and How They Know It*; Richard R. Nelson *Technology, Institutions, and Economic Growth* ("On the Nature and Evolution of Human Know-how" 115–138); and Carl Mitcham *Thinking Through Technology: The Path Between Engineering and Philosophy*. There are also plenty of examples of the discussion of evolved

mismatch between biology and ideas or artefacts, but I hope that this impasse can be avoided from the start: there is no one-to-one match between the biological encoding of successful forms in organisms and the mechanism that I will suggest for the evolution of artefacts below. However, the description of evolution that I have outlined, stripped down to its elements, continues to hold well in the case of the adaptations of material equipment. We need not get bogged down in the minutiae of how DNA encodes particular responses to the environment, for instance, and whether this has a corollary in artefacts; if the three criteria hold for an identifiable "organism" and an identifiable environment, and produce adaptive results, then I would suggest that the term "evolution" is a productive description of events.

---

artefacts: Karl Marx, for instance, described in *Capital* Darwin's then new theories as a way of explaining advances in technology as the work of the many in small improvements, rather than grand ideas coming from a talented few. Tim Ingold notes that "artefacts, too, may be grown, and that in this sense they are not so very different from living organisms...Just as the form of the organism is not prefigured genetically but arises through a process of growth within a morphogenetic field, so the form of the artefact is not prefigured culturally but arises through the unfolding of a field of forces that cuts across its developing interface with the environment" (*The Perception of the Environment*, 290) and this idea is similar to what I want to outline in this final chapter. W. Brian Arthur's *The Nature of Technology* and George Basalla's *The Evolution of Technology* are both concerned primarily with the invention and increase in complexity of artefacts. Both writers assert the distinction of their described mechanisms from Darwinian evolution, although they both also accept that users cause Darwinian selective pressures once the artefacts have been created. Bernard Stiegler, in *Technics and Time*, also presents a chapter on "Theories of Technological Evolution" (29–81). Stiegler is heavily influenced by the work of André Leroi-Gourhan, particularly *L'homme et la Matière* and *Milieu et Techniques*. Stiegler finds appeal in Leroi-Gourhan's theory of a "zootechnological determinism," that as the fish "must" evolve towards the amphibian so the knapped flint "must" evolve towards the steel sword (*L'homme et la Matière* 13); there are only a few paths evolution can take, and it operates within certain constraints. There is evidence for this idea and, as we'll see, wood and hands guide the shape of saws, arms and jungle the length of knives – such constraints certainly shape evolution and dictate impossible paths. But I dislike the notion that we might be able to predict evolutionary processes, or that any path might have been certain: evolution, to be evolution, must be an emergent process. Lastly, Stephen Shennan's collected volume *Pattern and Process in Cultural Evolution* offers a wide range of discussion of the potential corollaries between biological and artefactual evolution, and John Ziman, in *Technological Innovation as an Evolutionary Process*, offers an introductory chapter of the same title that includes a good survey of the range of (and challenges to) evolutionary approaches to material culture.

Firstly, then, we need to define the individual (for the sake of clarity I'll abandon the term "organism"): the codex.[7] Next we need to define the environment that places selective pressures upon it. With non-artefact entities the environment is easily defined: everything in the milieu in which it exists which can directly or indirectly impact upon its epi- or ontogenetic development and its surviving long enough to reproduce. But for artefacts, and for codices specifically, that milieu is more specifically defined, though the same principles apply: *we* are the defining selective forces for our artefacts – human users are, or (more accurately) *human use is*, the environment for our material equipment. Yes, conditions of, for instance, humidity do fluctuate, and this can have a profound impact on the survivability of a codex, but whether or not it gets to reproduce (or be reproduced in response to this organic environmental disturbance) is based on *our* allowing it to occur. Just as the vegetable matter's rising above the average neck length of our example animal was a way of the environment's "choosing" whether it got to reproduce, so we choose the fittest codex forms to be repeated, the ones that fit their environment, the ones that fit to us and to how we use them. That we can make this as a conscious decision rather than dispassionately enacting selective forces is irrelevant (and sometimes false); the end result is that the environment does or does not allow a new generation based on what has come before to emerge.

So if the individual is the book, and our use is the environment, then we have at least the arena for evolution to occur. Next the three phases of the evolutionary process need to be established.

5.1.2.1 Variation

Is there the capacity for mutation in the codex form? I think that we can clearly say "yes": from Gutenberg's first huge bibles to hotel pocket Gideons there are a range of sizes. From early metal codices and bound velum to contemporary recycled paperbacks there are a range of materials. From press, to lithography, to laser there is a range in print. From handwritten and illuminated manuscripts to Times New Roman there is a range of typography. In fact, for every descriptor of the codex there is a range of variation that shrinks and expands over time. It is also clear that there are predominant limits for this variation to take place

---

[7] Again, this is the codex of the popular imagination. Codices are clearly multifarious things, with a huge variety of forms being included under that name over the history of writing (and, more specifically, in histories of that history). For the purposes of this discussion, however, the term "codex" will largely continue to refer to a generic printed and spine-bound mass-market paperback.

in, with odd mutations (e.g., oversize atlases, intricate bindings, die-cut pages, etc.) extending or altering those standards, and emergent stabilities (such as the modern paperback, coffee table art collection, or dust-jacketed textbook).

5.1.2.2 *Selection*

For evolution to occur, pressures from the environment must cause a match or mismatch of fit for the individual. Let's use oversize books as the example: most codices are a certain size because we, as producers of selective forces, tend to not buy or to use or to demand books that we cannot easily hold or carry around with us. When there is no demand for an item, market forces (which, as they are related to our reception and deployment of books in a capitalist system, are part of structuring the environment for a codex, perhaps something akin to weather patterns in the non-artefactual realm) tend towards ceasing the production of that item. The proliferation and success of the average paperback book is based on a complex of such selective pressures over generations of produced texts which increasingly came to match the needs of our bodies in action amidst their own cultural and environmental milieu.

5.1.2.3 *Reproduction*

Lastly, is there a way for the attributes of existing codices to be passed on to the codices that follow them? Is there a relationship between "parent" and "offspring?" This would require us to think of the codices that follow those currently produced as being their "children." Or perhaps not. If we think of the collected group of codices as a "gene pool,[8]" then it seems valid to see any individual codex as the phenotypic expression of a set of possibilities that emerged directly out of that pool, and with the success of each particular expression contributing to maintaining the pool as it is.[9] The significant difference here between organism and codex seems to be one of agency: the organism struggles against the environment and *tries* to reproduce, whereas the codex is innately passive. But what *causes* variation, selection, and reproduction is not (or need not be) the measure of whether a process is evolutionary. In Marshall McLuhan's

---

[8] That is, the sum total of all of the living genotypes of a species.

[9] This is allied with Darwin's insight that it is not individuals, but species that evolve – an individual's success or failure has very little impact on its nature as an individual, but the thousands, millions, and billions of successes and failures of the collective are what shapes the species and the available potential ranges of expression.

terms, we are "the sex organs of the machine world, as the bee of the plant world, enabling it to fecundate and to evolve ever new forms" (*Understanding Media* 56).

We can extend these ideas out to any artefact. By shifting towards thinking of our actions with equipment as being able to exert conscious or unconscious environmental pressures we can see any number of ways that artefacts iterate over time. Think of the progression of hammers from rocks of all sizes held in the hands of early hominids to the now common form of a piece of specifically shaped metal on a wooden shaft. We can imagine all of the variations that occurred along that lineage, and the ranges of expression that still exist today, from small household ball-peen hammers, to lump hammers, mallets, and sledgehammers, and all emerging from the same selective and reproductive methods as those detailed for codices. For every artefact the same process must have occurred; if there are elements to be varied, then human users have and continue to apply selective pressures that will affect the manifestation of the next generation of the species. Like any other environment, we inform the species of our artefacts, cause them to take on new data, and this leads us back to knowledge – evolution is all about repeatable successful action.

5.2 Evolutionary epistemology

My reason for wanting to establish an evolutionary model for artefacts is that it opens up a branch of Philosophy that enables us to return to our discussions of expertise and a flat, posthuman approach to knowledge. If artefacts iterate through a series of evolutionary processes, then we may be able to consider them in terms of evolutionary epistemology (hereafter EE).

Epistemology is blended in EE with evolutionary theory to describe two increasingly distinct fields: (i) evolutionary principles applied to the progress of knowledge, particularly in the sciences[10] and (ii) the study of knowledge acquisition in living beings, where cognition and knowing are seen as evolutionary adaptations, and bodies, and the minds that they play a role in producing, are seen to reflect aspects of the world.

---

[10] See, for example, Franz M. Wuketits (ed.) *Concepts and Approaches in Evolutionary Epistemology: Towards an Evolutionary Theory of Knowledge* (in particular Erhard Oeser "The Evolution of the Scientific Mind"); Donald T. Campbell "Selection Theory and the Sociology of Scientific Validity"; and Louis Boon "Variation and Selection: Scientific Progress Without Rationality."

The first branch is easier to explain in basic terms due to the approach that we've already taken here. Extending the three phases of evolutionary process onto thought, this conception of EE often suggests that there is a unit of selection in knowledge, possibly the meme,[11] which can be varied, selected for, and reproduced. This is another example of Universal Darwinism, and though the field often attempts to explain subtleties such as the precise method of encoding information for the next generation, or the validity of making evolutionary claims about knowledge acquisition, the fundamental underlying mechanism is something that we've already addressed. Most theories of evolutionary approaches to technology fall under some variation of this first branch of EE.

The second branch is, if not necessarily more complicated, then perhaps less intuitive. It also rests on the fundamental principles of evolution, but its claim is not, or not solely, about the mechanism of human knowledge acquisition and transmission. This second approach sees human knowledge as an adaptation, and, in some instances, that every evolved adaptation in an organism is best thought of as *being* knowledge. I'd like to adopt this latter vision of EE as outlined by Henry Plotkin in *Darwin Machines and the Nature of Knowledge* as the elegance of his theory coheres with the bare-bones mechanism of evolution that enables Universal Darwinism to function. I would like to argue that Plotkin's approach allows us, through its stripping out of the specificities of biological organisms' reproduction and genetic encoding of information, to talk about EE as it might apply to artefacts in their alternative

---

[11] Richard Dawkins coined the term "meme" in chapter 11 of *The Selfish Gene*, "Memes: The New Replicators." For Dawkins, a meme is the smallest unit of meaning spread via cultural phenomena. In his original formulation it is a relatively unsophisticated idea, a rough analogy to the gene for something which might be replicated in a purely cultural environment: "Examples of memes are tunes, ideas, catch-phrases, clothes fashions, ways of making pots or building arches... [M]emes propagate themselves in the meme pool by leaping from brain to brain via a process which, in the broad sense, can be called imitation" (192). The most significant fleshing out of Dawkins' idea is Susan Blackmore's *The Meme Machine*, but the term has also achieved pop cultural standing, becoming a meme of its own, describing rapidly spreading (and equally rapidly dissolving) inexplicably popular media events, small quirks (a particular photo, a turn of phrase, a way of acting) which suddenly seem to be everywhere (or everywhere within a subculture), acting as satire, taking on semantic weight, or simply provoking joy or ire.

environment of human use. This, in turn, can enable us to fully realise a posthuman conception of our artefacts' knowledge.

5.2.1 What the stick insect knows

Plotkin's central idea is that as organisms adapt to their environments via evolutionary selective pressures, what successful individuals pass on to their offspring in each generation is not just genetic instructions for building new bodies within set ranges, but knowledge about the world that came before them (hence EE). His most striking example is that of the stick insect: a stick insect looks like a stick not because it tries or learns to, but because generations of its ancestors survived more frequently the more that they looked like sticks, avoiding predators long enough to reproduce and to pass on genes which stipulated increasingly stick-like bodies. Plotkin argues that the stick insect's body therefore possesses knowledge[12] of an aspect of the world far greater than its own mind is capable of.[13]

This immediately raises the question "why use the word 'knowledge' to describe an adaptation?", and Plotkin asks the question himself: "why take the further step of equating adaptations with knowledge?... How can the wing markings of a moth[, for instance,] be knowledge?" (117). He then defends his word choice on grounds that should be very familiar: "knowledge, in its most common meaning, denotes a mental state that bears a specific relationship to some features of the world" (4). For Plotkin, when we say that humans "know" something we're stating that there is parity between two things: "a brain state, which is a part of organismic organization, and the world itself... which is the feature of environmental order relative to which that brain state stands" (117). As

---

[12] I'll go on to clarify this under the ideas of knowledge and information that I outlined in the last chapter, but to be clear from the start: when Plotkin talks about "knowledge" I tend to think of it, instead, as "information" until it is tested.

[13] There is a relationship here to an idea explored by Karl Popper in *Objective Knowledge* – in this work Popper is also interested in variations on the possibility of knowledge without a "knowing subject":

There is no sense organ in which anticipatory theories are not genetically incorporated. The eye of a cat reacts in distinct ways to a number of typical situations... [and] these correspond to the biologically most important situations between which it has to distinguish. Thus the disposition to distinguish between these situations is built into the sense organ, and with it the *theory that these, and only these, are the relevant situations for whose distinction the eye is to be used.*" (71–72, emphases in original)

I argued in Chapter 4, there must be a brain state that bears some partial correspondence to the thing in the world:

> Knowledge is always something that comes in two parts. There is the "knower's end" of knowledge, comprising feelings, brain states and ... the means of expressing the knowledge; and there is the "world's end" of knowledge, which is that aspect of the world that is known. All knowledge is a relationship between the knower and the known. (10–11)

Again, this is not to say that there's a miniature version of the world playing out within our brains, simply that for an act of knowing to occur there must be a state of cognition or of memory which has a physical instantiation and which maps to our experience of, recall of, or interaction with an object or state in the world, and in that sense alone we might be said to incorporate or represent it. Though Plotkin doesn't take an object-oriented approach, his outlook is wholly compatible with our discussion so far with three adjustments: first, that the embodied knowing that he describes is the product of being informed by sensual rather than real objects in the world. Second, that knowledge isn't always held in the bodies or brains of single things, but can instead be spread across action in the world as individuals enter into systems (though this latter tends to matter far less for the kinds of non-human relationship that we're discussing here). And third, that what is coded during evolution is information-data, data "in response to" – in the terms that we have discussed, it is only knowledge if it provides the potential for repeatable successful action due to its coherence with reality limiting the flow of information from the world. This, however, is largely what Darwinian evolution is all about: causing organisms to better fit their environment in order to be able to act most successfully.

By envisioning knowledge in the fundamental and pragmatic fashion outlined above, Plotkin is able to use the word to describe adapted biological organisms' relation with the world. Evolutionary adaptations, like all kinds of knowledge, always have a

> relational quality. Every adaptation comprises organization of an organism relative to some feature of environmental order ... The wing markings of a moth stand in relation to the nervous system of a predator, specifically the way in which that nervous system is wired such that the "eye" [of the moth's markings] startles the predator and perhaps causes it to flee ... All human knowledge has the same two-component relationship that adaptations have. (116–117)

This matching of bodily- or cognitive-states to world-states is the underpinning assertion for Plotkin's vision of EE and lends further support to a flat approach to knowing which can include non-human artefacts.

Adaptations conform the bodies of the evolved organism to the worlds that housed its lineage; a stick insect's body is the sum total of the information gleaned from the environments of its ancestors combined with data prompted by its own experience of its present environment during epi- and ontogenesis. In this way, stick insects came to incorporate a knowledge of the encountered environment into their being – their data allowed them to act successfully. Simply put, they are partly material instantiations of the information that the environments that preceded them favoured insects that looked like sticks meaning that their bodies continue to know how to act in their present milieu. As the philosopher Tommi Vehkavaara describes it:

> When natural selection makes changes in genetic information, this is interpreted [in this branch of EE] to be a knowledge process, and if this change is adaptive, it means the increase of knowledge ... Adaptation through natural selection can be seen as ... *evolutionary learning*, where lineages, populations, or species (but *not* individual organisms) are considered as individuals that are learning about the conditions of the survival of their "members" – these "collective individuals" are "testing" their environments by their "individual individuals". These supraindividual learning processes explain how individual organisms have got ... "*a priori* knowledge." (208–209, emphasis in original)

Vehkavaara's description of the idea underpinning this branch of EE makes the relationship between adaptation and human knowledge explicit: if we think, as Darwin did, of the species as a whole as an individual, then it becomes clear how this process is akin to a more traditional definition of knowledge – the species pushes at the environment, learning by sacrificing the individuals of which it is comprised to selective forces, like turning the tumblers of a combination lock, winnowing number sequences until the code is known and one state bears a successful relation to the other. The *species* is informed (as I have defined information) resulting in its next generation of individuals possessing knowledge about the world from the moment that they are born.[14]

---

[14] Each individual is also informed during its lifetime, but the only thing that has an effect on the data in the gene pool for the next generation is its "fitness," its fit with the world it finds itself in and, crucially, its ability to reproduce the data that underpins such fitness.

To reiterate the argument of Chapter 4, when we say that we, as humans, know a stick insect looks like a stick, and that this allows it to avoid becoming prey, our brains are performing a variation of the stick insect's evolution: conforming to an aspect of the world as it is encountered, materially encoding data, spinning the tumblers until the knowledge fits, until we can be repeatably successful and slow the flow of information. In fact the human ability to apprehend and reflect upon the world, such that we can make claims like "I know X," is itself an adaptation, a manifestation of the information that the environments that housed our ancestors favoured those individuals who were able to acquire conscious knowledge faster than the pace of embodied generations. This leads Plotkin to state that "[i]f adaptations are knowledge, and if what we commonly call knowledge (or better, our ability to gain knowledge) is an adaptation, then what in ordinary everyday life we call knowledge is actually a special form of [the] much wider phenomenon... [of] biological knowledge" (xvi). This again supports the approach to knowledge that I took in Chapter 4, though by making the distinction between information and knowledge we are able to usefully nuance the idea.

5.2.2 The Machete's evolved knowledge

The question now becomes whether, if the development of artefacts can be seen as an evolutionary process, the theories of this branch of EE might also be applied to them; are artefacts instantiations of information about the environments that shaped their lineage, data that manifest as knowledge about the milieu that they now find themselves in? For our stick insect, her ancestors were a mix of more or less stick-like insects; the fact that she exists today, to over-simplify, shows that her ancestors were the most stick-like. The gene pool of early insects generated billions of more or less "stick-y" bugs over time through various mutations, and those that most resembled sticks, who were better camouflaged, and passed on their stick-like natures to their offspring resulted in our current individual – she is an instantiation of the sum of the biological information coded into her by the environment informing her species via the bodies of her ancestral lineage. If we want to say that material equipment can be a similar instantiation, then we must find some parity between an artefact and these aspects of the stick insect.

The modern machete is a ubiquitous tool in many tropical countries where it is used to cut away vegetation when travelling through dense jungle, to harvest tough crops such as sugar cane, and also for butchering practices where a cleaver is a common alternative in other parts of the world. It's essentially a long knife, around a third to half a metre long, often with a slightly curved blade that's typically set into a wooden or plastic two-part haft bolted together through a full tang. The machete, for our discussion of EE, is the artefact (the non-biological corollary of the organism), and its community of potential users, and their selective pressures, are its environment.

Variation in knife manufacture and design, as with codices, is clear: from the first stone blades used by early hominids through to multi-component contemporary cutting tools, the sheer variety of blade lengths and shapes, handle styles, materials, components, etc. is staggering. This is to be expected of a tool that has been put to so many different uses around the world and for so long; every human culture has found the need for a sharp edge. Each new development comes from a mutation derived from environmental effects that alter the range of a feature to be expressed. For example, in a culture where knife blades are typically between two and five inches, a 15 inch blade is a mutation which, if used successfully, permanently alters the potential set of blade lengths for future generations of knives.

In this way, the contemporary, relatively standardised machete comes from a process of **selection** dictated by its environment. The stick insect's ancestors ran the risk of being eaten if they were not significantly stick-like; for every mutation which made them more vulnerable, predators, a pressure of their environment, frequently stopped them living long enough to reproduce. But for every mutation that made their genes more likely to provide a range of form and colouring akin to their surroundings, a range of appearances more accurately fitting the vectors of a stick, then the environment "rewarded" that trait by allowing it to be passed on. The insects' bodies matched a system in the world that remained relatively consistent over generations, thus their biological information about and knowledge of that aspect of the world grew – they had incorporated a highly-coordinated dataset into themselves, and more specifically into the species-as-individual. Machetes clearly don't look like any aspect of their human environment, but I would argue that the same process of selection occurs. Fitness to the environment for artefacts is the same as for organisms in as much as success depends on matching aspects of the environment in such a way that that environment doesn't obliterate the traits of the particular instantiation. In a tropical climate the machete shape is the best fit for its environment. This is not to say that the machete matches the jungle, or incorporates an aspect of the jungle, it doesn't; the machete has no evolved knowledge of jungle environments. But it does have a knowledge of how part of its environment *intersects* with the jungle, how its human users experience that terrain. A short stone knife is no use for swiftly clearing vegetation, so when metal came along, which allowed for thin, strong blades that could be carried easily, it was adopted for that task (and simultaneously for many others). Metal also introduced the possibility for a new variable: blade length. A longer blade allowed for large slashing motions to be made – inefficient for precision work, but perfectly suited to human passage through tropical terrain. *This* is what the machete matches: the repeated (postphenomenological) moments of action where knife users meet (and met) the jungle. This is the aspect of their environment that relates to blade length just as the individual stick insect is the product of past insects' repeated intersections with predators that are unable to distinguish between sticks and insects. Blades would have become longer and longer as users discarded shorter blades and created, or requested the creation of, increasingly machete-like knives; similarly, blades that were too long or unwieldy would have quickly disappeared as failed experiments, failed mutations.

And this is the moment of reproduction. Stick insects, having successfully evaded the selective pressures of their environment, would mate and return their particular combination of genes to the gene-pool, causing new phenotypes to express them in new ways, more or less successfully. The machete doesn't have genes, and it can't directly facilitate the creation of the next generation of long bladed knives, but the third evolutionary criteria still stands. When the stick insect mates this is also an adapted behaviour and therefore an instance of information and knowledge. The ability to mate relies on knowing that there will be other stick insects in the environment with which mating can occur. When offspring are produced, an aspect of the environment (another stick insect in this case) has caused the organism's genetic material to be reproduced; in this way reproduction is a concert of individual and environment. Machetes have information about the forces of their environment of users: their traits also get reproduced when an aspect of the environment causes it to occur, that is, when a long bladed knife is used successfully an individual user is more likely to recommend it to other potential users, and to produce or request this trait themselves when they next need the tool. Thus a machete's blade holds information not only about how humans encounter jungle plants, or any other activity in which a blade is put to use, but also of the consumer forces that can allow such a blade to come into being and be repeated. Blade length is a heritable trait in knives.

In his discussion of EE, Vehkavaara reiterates the pragmatic approach to knowledge that Plotkin and I also share:

> Discursive linguistically expressed justification is not always necessary – if the ability to act *is* (successfully) *demonstrated*, no argument can overcome this ultimate proof of knowledge. This kind of demonstrable knowledge connects us to other forms of life – every living creature needs at least some knowledge how to act successfully (in its environment). Of course, knowledge does not determine the action it enables, it is just the precondition for the action. Although an action can be seen as a presentation of knowledge, the actual action is not necessary for the existence of knowledge – knowledge is *potential action, the power to do*. (210, emphasis in original)

Vehkavaara emphasises the potential for successful action as underpinning what it means to know, but does so in the context of Plotkin's biological knowledge, and this further shores up and draws together our discussion over the last two chapters. A stick insect is put into action in the act of being a stick insect; a genotype is knowledge-data when it has *potential*, potential that is realised in the actions of every living epi- and ontogenetically produced phenotype. Every second that the stick insect is alive it demonstrates that its body has information about and a knowledge of the consistent aspects of the environments that led to its being, that there will likely be oxygen to breathe, food to eat, light to see by, predators to evade, and other stick insects to mate with.[15] The machete differs in that it doesn't act second by second, it only acts during use, but really the stick insect also only ever puts its biological knowledge into action when it is in concert with its environment, it just happens to never be outside of that environment, or something approximating it. A stick insect born into a vacuum doesn't act; it has nothing to know and simply ceases. A machete outside of use also doesn't act; it rests on the table, hangs from a hook, or is attached to its owner's belt, and cannot demonstrate and therefore cannot prove its particular knowledge until that concert with its environment begins. But the moment that it is picked up it comes into action, and the ensuing success of its use is a measure of the quality of its inherited information, of the accuracy of its embodied knowledge. As David Rothenberg notes "[n]o machine stands apart from its creator, no tool makes sense outside of its use" (xv).

---

[15] Just to reaffirm the OOO position, what it knows is these aspects as it encounters them, as sensual objects, rather than satisfying their reality.

Artefacts exist as potential knowledge until they are seen in action; they have, in Vehkavaara's terms, "the power to do."

5.3 Coda

Were we to say, simply, that the stick insect knows her environment, we would be wrong. Instead, her body possesses data from her species' being informed by past environments such that she is able to act in concert with her new milieu. But there are also always elements of even those specific things that she's meant to correlate with that escape her; her body, in object-oriented terms, doesn't ever satisfy the world, in fact it only relates to a tiny facet of the environment whilst the rest recedes; even her bodily information about sticks is vanishingly discrete, reducing them to a set of qualities as encountered by a predator's eye. I chose the machete as a corollary example because, as with our stick insect, we can reduce the discussion to what is functionally a single variable, comparing a range of blade length to a range of stick-like insect forms. In this way, as the stick insect's stick-iness stands-in for any single organismic adaptation, blade length in the machete is made to stand-in for any one variable of any artefact, from the weight of a hammer to the size of a silicon chip. With this in mind, and to conclude the discussion, I want to return our attention to the codex and the e-reader to see how the discussion presented here impacts on our understanding of the former feeling somehow superior to the latter, taking us full circle.

A codex body is made up of myriad variables: mutations in form and size, bindings, prices, printing speeds, positioning of marginalia, materials, typography, variations of all parameters that have been generated, selected for, and reproduced in successive generations by the environment of potential creators and consumers. Our ancestors ensured, as we continue to do, that these media artefacts became fitter, fitted to us, adapted. And if all adaptations represent information about or knowledge of how to act in an environment, then that means that books do not just contain knowledge, but that they *are* knowledge, they *are* information, and they manifest the history of their ancestor's interactions with the humans that held them. I realise that this seems to suggest a teleological progression towards perfection, but of course this is far from the case, in this instance or under any evolutionary mechanism. This apparent problem is solved by recognising the flux in any environment: sometimes codices are selected primarily on size, sometimes on affordability, sometimes on perception of exclusivity. With these and a thousand, a million more selective pressures, many in competition, we can see how a form can settle into a loosely fluctuating aggregate of desire rather than a simple perfection – the environment will inevitably throw off this current stability again, and e-reading presents exactly the kind of force that might make this occur. In biological terms we could call this "punctuated equilibrium," the tendency for species to find and maintain certain successful forms before an alteration in the environment prompts a period of rapid change until a new balance is established.

We may feel uncomfortable with the idea that codices possess knowledge rather than simply store it, but this only reflects a prejudice surrounding the use of the word "knowledge" as a narrow band of human experience, rather than seeing that band as a subset of something that we share with the world that we have emerged from and remain in complex concert with. When I say that a stick insect has a knowledge of the world within its body, I suspect that its own ability to cognise to some extent takes on part of the weight of this unintuitive assertion. An artefact, a knife, or a codex, however, cannot allow us to displace that same weight of a knowledge assertion onto its own cognisance. But as we've seen, the argument is not to say that printed books are somehow conscious of their acquirements, any more than stick insects choose to reflect an aspect of their milieu, machetes choose to cut through jungle, or our own hands are aware of their frequently conforming to the objects that we would like to grip. In each instance a bundle of adaptations represent an approximate transcription of a tumultuous past in physical form; in every printed book's body there is speck of the use of every printed book that came before it.

And maybe *this* is why e-readers can feel so wrong: bound books have been informed by us, by a hard fought struggle, but our new artefacts, though they draw on some of the same data as the codex, they may seem for some readers to be back with the typographic amoeba, under-developed, under-evolved. When I initially defined technology in Chapter 2, I came at it from the perspective of the individual user encountering a class of objects, and the definition reflected this. But now we can consider it from the *artefact's* perspective, where the group-as-individual's growing knowledge of its environment might be just as important. If the individual user doesn't feel that an artefact is sufficiently well fitted to them, then the move from device to technology will always be impeded; an unwieldy thing, an overlong machete for instance, will never be incorporated, will always slip from readiness-to-hand. Part of what e-readers need, or have needed, in order to facilitate an increasingly regular move from device to technology, is time and generations. Contrary to the polarising folk phenomenological reports that we've seen throughout this work, I would argue that even the most profoundly unintuitive equipment or method of performing a task can be normalised over time such that the practiced activity can supplant a mode which would have previously seemed preferable. Equally important, however, is that the equipment and the method for the practice itself meet some threshold of usability.

It may seem a simple point, that in order to use something it must be useable, and that we get there by choosing usable things, but the language used to discuss technology, the familiar tropes of such discourse that we have repeatedly seen, demonstrate that we routinely miss this idea. A drive to technologise something can seem initially insurmountable when we put new equipment to use in order to accomplish a task that it is barely informed of, use that it barely knows: early generations can be put to work by specialists and adopted by curious users before they conform and are conformed to, putting off the less hardy with the suggestion that it will never be for them.

But it is also not enough to simply say that the reader will get used to an artefact over time because every act of reading is a meeting of (at least) two evolved bodies, the reader and the equipment. Whilst users might increase their knowledge of how to engage with an e-reader, this means nothing if the artefact has no information about them.[16] Each reader is part of the environment in which each item of reading equipment sits, they are an aspect of the environment to be known, a little bit of selective pressure on the gene pool, but also on the individual equipment, like a successful predator being a part of the environment that affects all stick insects, but also, in that moment, a specific individual. And, just as with the individual stick insect meeting her individual predator, any artefact (such as an e-reader) meeting an aspect of its environment (such as a resistant reader) can pass on the information it gains from that meeting to the next generation of its species. For the stick insects this can make the species come one step closer to looking more like sticks: the individual

---

[16] Note that the evolved knowledge of the artefact has not evolved in response specifically to the current single reader, but to an environment of users. This need not trouble us; knowledge, as described here, is a measure of fit to action in the world however that might come about. The stick insect has inherited information from past environments, and yet we can say that it has a knowledge of the present because aspects persist with which it has a successful relation. The same is true of the e-reader; any feature of it, any data quality, which matches us specifically (or at least the aspects of us that it encounters) is knowledge. The growth of knowledge isn't always an evolutionary process, but evolutionary processes do tend to produce knowledge.

just didn't cut it and their genotype is eliminated along with their phenotypic expression, or they made it through this time giving them a chance to go and find a mate. For every successful event, that stick insect is informed that it lives in a world where its particular body is knowledge, that it will allow it to succeed. In the same way, every time a resistant reader puts down a Kindle and says "no, this just isn't for me," and never buys one, and never recommends one, and maybe even actively tries to discourage other potential readers from getting one, a similar act of information occurs: *that* Kindle didn't exist in a world where it could act successfully with *that* reader, with *that* aspect of its environment (success, for the Kindle as an artefact, being use, being put into action). This affects the whole species, even if in a minute way, and minute effects played out over and over again is what evolution rests upon. Thanks to the peculiarities of their environment, artefacts tend to experience a much more focussed (as goal-directed) set of selective pressures on each generation. Artefact evolution can therefore be swift, at least until its knowledge of the environment is such that most of its encounters are successful and less information tends to flow. The codex has reached a relative equilibrium; the artefacts of electronic reading are still under pressure.

With e-readers, indeed with any artefact, there is a hugely complex interplay of individuals, groups-as-individuals, and environments (which include or are comprised of individuals).[17] For e-readers we have (at least[18]):

- The individual e-reader in its environment of pressures from readers.
- The individual e-reader's encounter with an aspect of its environment: an individual reader.
- The group-as-individual of e-readers, the species, experiencing the combined selective pressures of its environment of readers.
- The individual reader in their environment that includes e-readers.
- The individual reader encountering an aspect of their environment: the individual e-reader.
- The individual reader contemplating the group-as-individual of e-readers.
- The group-as-individual of humans, not just the literate and sufficiently affluent and inclined members of the species, experiencing the combined selective pressures of its environment that includes, for some members, e-readers.

---

[17] There's a nice parity here with Bogost's discussion of OOO seen above, where objects exist as individual things, in relations with other objects, and as relations of objects.

[18] And to limit it to fairly direct relations.

These kinds of encounters are at play in the issue of getting used to any new equipment for reading, or to any artefact at all. To bring the language of the last few chapters together on this complexity, we could say that when the happy e-book reader tells the reluctant screen reader "you'll get used to it," all they can mean is: "when my knowledge of the object became strong enough, the whole-composite of sensual aspects that I encountered correlated with, but still hid a real object that also had a good enough inherited knowledge of aspects of me such that a reading experience could occur that I would deem successful. I've become informed; I'm now rarely surprised by e-reading enough to disrupt the technological engagement." This highly awkward phrasing (as ever, it's far easier to say the sun rises …) clearly has no bearing on whether the same will be true for the reluctant screen reader; it's only a statement that it is possible.

But, following on from this, the accumulated pressures of all the happy and reluctant readers have an effect on the *next* generation of e-reading equipment. The next generation will tend to have a better knowledge of its environment, and we can see this in the flattening and thinning, speeding up and brightening, cheapening and simplifying of the generations of Kindles and iPads. Over time a feedback loop of knowledge can emerge. The stick insects, generation by generation, look more and more like sticks in an environment which includes individual predators that are either a) increasingly sensitive to perceiving the distinction between sticks and insects (resulting in even more stick-like insects in the next generation) or b) have gone off looking for other food (in which case the stick insects thrive until some mutation throws this new balance off). For e-readers, such a feedback loop will either (a) make each generation of e-readers better suited to increasing numbers of individual readers who are, in turn, better equipped to interact with them, or (b) there will always be a significant amount of readers who feel that, no matter how hard they try, they won't overcome the gap in knowledge instantiated by the devices, and their discontent will manifest itself in keeping e-reading as a minority pursuit, or one with a significant amount of detractors, whilst simultaneously throwing up all sorts of strange mutations in e-reading devices as they try to adapt.[19]

---

[19] Note that when I say "try" here I'm not suggesting that e-readers realise their problems and attempt to improve upon them. I merely want to express that their relationship with the environment, as with the stick insect finding a mate prompts the next generation into being, and when a stable survival/success strategy is not established there is a tendency towards diversity in the ensuing offspring as all sorts of traits appear survivable; there is no accurate established knowledge.

This, I would suggest, is a broad mechanism for rapid change in artefacts, for getting used to things and developing expertise, for the kind of exponential growth that can lead from our roughly dividing carcasses with chipped stones found in our surrounding area, to moving through the undergrowth with the practiced swings of a machete blade which feels like an extension of the arm rather than a tool in use. But whereas a decent knife can last for a lifetime, the vagaries of market growth, planned obsolescence, and manufactured consumer demand have led to an environment for new equipment where replication is increasingly rapid and responsive. This can be positive in some respects, with the right tools for important jobs, medical equipment for example, going from idea to indispensable technology within a couple of generations. But there is also a homogenising influence where mutations aren't given a chance to shine as users take little time to comprehend them, and the pressures of what we think we want march over what we may be better off with. Such problems are largely not under consideration here, but I hope that the language outlined is applicable to such concerns. I don't believe that e-reading is an inherently bad thing, I do think that we can adjust to it, and that its equipment will adjust to us, but I also fear landfills full of misunderstood artefacts and a culture where new mutations are prized over what could be most successful. As with all environments, we can be cruel and demanding, but when an individual finds its niche we also have a history of ensuring that it flourishes; we tend to become better for it.

Conclusion

Judith Donath, the founder of the Sociable Media Group at MIT's Media Lab, contributed a chapter about a car that she used to drive to Sherry Turkle's collection on *Evocative Objects*. Donath describes the ways in which her Ford Falcon seemed to have a life of its own:

> I negotiated with the Falcon when it exhibited its own preferences for speed, for direction. It required work to anticipate its likely reaction to my wishes. If I wanted to go faster I could not just depress the accelerator, because it would likely react by stalling. Instead, I needed to slowly add pressure to the gas pedal, letting up when it felt resistant and listening for a shifting of gears...The Falcon stalled in the rain, and as it grew older, it stalled on any damp, low-pressure day, teaching me to become, like the pilot of a small plane, attuned to the meteorology of frontal systems. (157)

The machine pushed back, requiring Donath to take on a new range of expertise in order to use it effectively. Sometimes it worked, but sometimes the car was elusive; it, and the systems that it interacted with independent of its driver, got away from her best efforts. Now Donath drives a BMW: "It is, as they say, 'the ultimate driving machine.' It does what I say, I feel no need to negotiate with it; I do not feel that it has a will of its own...New, it is a commodity with nothing to distinguish it from the many other BMWs in this city. Its biography is yet to be written" (157). The BMW seems to have a far greater knowledge of Donath than the Falcon, evidenced by their repeatable successful actions together, and this makes her task easier – the machine doesn't push back, doesn't require a readjustment of information; expertise comes more easily and remains stable due to a designed coherence. But there is a wistfulness in

Donath's account; she is waiting for the artefact to be written upon by the world so that it takes on a greater life of its own, like the Falcon. We often like our most treasured artefacts to resist us a little, to have their quirks that we must accommodate – in their unpredictability it makes them feel more human, or at least more alive.

I've tried, throughout this work, to understand the role and reality of artefacts and technologies better, to resist simplifying what must be amongst the most complex and fundamental of our interactions with our environment, and to query the established ways of speaking about these encounters. The existing discourses often don't seem to do justice to technologies, either seeing them as unnatural or alien, or mundane and uninteresting. They are none of these things; they are deeply coordinated with what it means to be human and possess a richness in their being and in their phenomenal appearance that shouldn't be neglected.

The posthuman view of postphenomenological encounters, and of knowledge and information, that have been established over the course of this book leave us with a strange world, one where every thing can be seen as informing every other thing that it meets, leaving readable and potentially actable data in each other's forms. And yet all things never meet one another as they are, always reducing and approximating, always escaping and surprising one another. The human use of technologies, however, gives an example of how humans draw, maybe more than any other creature, consciously closer to the real world that they find themselves in, a world that, as Nabakov noted, must hang spectrally around us even as we cannot even fully know ourselves – the shell in the ghost. Our expertise, with our e-readers, our hammers, our bodies, must draw us into some coherence with the real objects that escape behind the sensual tools that we interact with, and this offers a model for how any sensual approximation of contact might come to occur. What makes us different from all other things is the extent to which we can also act on the data that we find bound up in our environments, and this is the basis for technological use, and it's what makes us special.

In the first chapter, I tried to show that the word "technology" necessarily fails us when it allows commentators to describe the use of equipment as being, de facto, "unnatural," and to suggest that the new equipment for reading on screen, if they can become a technology as defined here, are no more unnatural than the codices which came before them. E-reading is a potent example of something which we can see being turned from an unfamiliar device into a natural(ised) artefact by some readers, whilst just as many, maybe more, lament from the side-lines. In another generation I would have talked about the mobile phone, the home computer, the car,[1] but I think that e-reading remains the best possible example today for talking about an encompassing definition of technology because it affects the written word, a technology in itself, and one which underpins almost all of our cultural existence in one way or another – e-reading unites embodiment, practice, and language; no wonder that we care so much.

The most persuasive argument for the superiority of the codex form, over any representation of digital script, typically lies in the realm of haptics and interfaces. It is unsurprising that a form that has evolved over hundreds of years should have naturally worked its way towards fitting our bodies exactly; the codex has been tailored, adapted by the repeated use of generations. And the haptics of the printed form do make sense: it fits the hand; the pages are thin for the most amount of storage in a compact and cheap space; it is more portable than a carving, and more adaptable than a scroll; less fragile than an iPad; less complicated than a Kindle. E-readers possess a different, and currently lesser, knowledge of their users than a codex does. But codices aren't perfect; no environment is static, stability is always broken, evolution always occurs. E-readers, or something like them, will lose their deviciveness and become mundane before becoming essential, and this is the path that every artefact we value or even love has gone through.

I cannot pretend that I have been able to tell the whole story of how we get used to new equipment. What makes people pick up a new device and try to make it a part of themselves? What motivates them to keep trying when it seems, unrelentingly, to remain a device? What finally makes that device match the user enough that it can be used effectively? Maybe these questions are unanswerable; if they could be predicted, then engineers would get rich by only producing equipment which suited users exactly, rather than throwing up iteration after iteration to see what sticks. What I hope I can say is that part of the reason that such questions are challenging is due to the hidden complexity of the interactions between individuals, groups-as-individuals, and environments of individuals. The environment that must be known alters itself, the individual who must know is changed, and the adoption of equipment that

---

[1] Examples of work that uses such totemic artefacts include: Lisa Gittelman *Scripts, Grooves, and Writing Machines: Representing Technology in the Edison Era*; William Boddy *New Media and Popular Imagination: Launching Radio, Television, and Digital Media in the United States*; and Friedrich Kittler *Gramophone, Film, Typewriter*.

is frequently technologised must represent, in some ways, a conforming between actants. The exact nature of every conformation is impossible to know; each is tied to the vastness of environmental influence and to the millions of individual users meeting millions of individual artefacts and reporting back, in various ways, to the pool which results in the next generation. But that such gains occur is written into the body of every object and into the mind of every expert user; in every repetition of a successful interaction there is the demonstration that knowledge of one kind or another has been gained.

Such a process of technologising is an act of making things feel right, of adapting to them, but also of clarifying, typifying, and setting up the boundaries of an artefact and what it can accomplish in a soft-assemblage with ourselves. To point at an object, or a set of objects, and call them "technologies" says nothing about the object as-it-is beyond our belief in its conforming to our bodies (and, eventually, our bodies to it) in such a way that it has the ability to provoke a relatively standardised, predictable, and repeatable sensory and use experience. In this process it can seem that we are getting to an artefact's essence, to how it is in itself. Therefore, when there is a fluctuation which challenges that illusion of access, it hits us all the more strongly. The fact remains that even in an expert technological, that is, deeply intimate, interaction, whilst we may catch an object's whatness in glimpses through coherence, all that we can truly rely on remains the equipment's "thatness," not what makes it what it is, but simply that it exists to us at this moment as this thing. This perhaps goes some way to explaining the profoundness of the response to a very familiar object malfunctioning. Imagine writing with a pencil which suddenly splits and pierces the skin of your writing hand; running in favourite shoes and feeling the tips give way as your toes push out of the fabric; the steering wheel of the car, even for a second, becoming unresponsive. These scenes all have an element of danger, but in those reflective moments following the event there is also an aspect of shock that the familiar, the "fully understood," has suddenly became alien; whatness is revealed as mere thatness, and simply understanding that something exists is no future guarantee of how it will perform.

But though the essence of the world always retreats, though there are always aspects that we will never have access to, we mustn't forget the reality of those things that we really do encounter, however much our limited and shaped perception is a requirement of their even coming into being. The sensual objects of our lives are beautiful monsters, born of an empirical world, and its strange and mediated collisions, and the asymptotic approach of knower and known.

Technological interactions are those that are undertaken with the tools that remodel us even as they invisibly assist in shaping the sensual environment that we encounter, and that enable us to act successfully within that lived environment despite its discrepancies with reality – for now, after doing my best to learn about technology, this is my new common-sense approach. I hope that it feels richer than a view which sees technology as just a means to an end and that it advocates for technology as being firmly within our nature. But, if nothing else, I'm happy with the claim that whatever knowledge I've committed to these pages, the pages themselves know better.

Bibliography

Ackerman, R. and M. Goldsmith, "Metacognitive Regulation of Text Learning: On Screen Versus on Paper." *Journal of Experimental Psychology: Applied* 17.1 (2011): 18–32. Print.
Ahrens, Christian. "Technological Innovations in Nineteenth-Century Instrument Making and Their Consequences." *Musical Quarterly* 80 (1996): 332–340. Print.
Amable, B., Elvire, G., and Stefano, P. "Changing French Capitalism: Political and Systemic Crises in France", *Journal of European Public Policy*, 19:8 (2012): 1168–1187.
Ambrose, Stanley H. "Paleolithic Technology and Human Evolution." *Science* 2.291 (2001): 1748–1753. Print.
Arthur, W. Brian. *The Nature of Technology: What It Is and How It Evolves*. London: Allen Lane, 2009. Print.
Austen, Jane. *Northanger Abbey*. Oxford: Oxford UP, 2003. Print.
Back, M. "The Reading Senses." *Digital Media Revisited: Theoretical and Conceptual Innovation in Digital Domains*. Eds Gunner Liestøl, Andrew Morrison and Terje Rasmussen. Cambridge (Mass.): MIT Press, 2004. 157–182. Print.
Baird, Davis. *Thing Knowledge*. Berkeley: U of California P, 2004. Print.
Barack, Lauren. "The Kindles Are Coming." *School Library Journal* 1 March 2001. Web. 18 May 2010.
Basalla, George. *The Evolution of Technology*. Cambridge: Cambridge UP, 1988. Print.
Bauerlein, Mark. *The Dumbest Generation: How the Digital Age Stupefies Young Americans and Jeopardizes Our Future (or, Don't Trust Anyone Under 30)*. New York: Penguin, 2009. Print.
Bearne, E. "Rethinking Literacy: Communication, Representation and Text." *Reading: Literacy and Language* 37.3 (2003): 98–103. Print.
Bekkering, H. and S. F. W. Neggers. "Visual Search Is Modulated by Action Intentions." *Psychological Science* 13 (2002): 370–374. Print.
Bell, Vaughan. "Don't Touch That Dial! A History of Media Technology Scares, From the Printing Press to Facebook." *Slate* 15 February 2010. Web. 18 February 2010.
Benzaquen, Adriana S. *Encounters with Wild Children: Temptation and Disappointment in the Study of Human Nature*. Quebec: McGill Queens UP, 2006. Print.
Berti, Annar and Francesca Frassinetti. "When Far Becomes Near: Remapping of Space by Tool Use." *Journal of Cognitive Neuroscience* 12.3 (2000): 415–420. Print.
Bijsterveld, Karin. "'A Servile Imitation': Disputes about Machines in Music, 1910–1930." *I Sing the Body Electric*, Ed. Hans-Joachim Braun. Wolke, 2000. 121–147. Print.
Birkerts, Sven. *The Gutenberg Elegies: The Fate of Reading in an Electronic Age*. Rev. ed. New York: Faber and Faber, 2006. Print.
Blackmore, Susan. *The Meme Machine*. Oxford: Oxford UP, 1999. Print.
Boddy, William. *New Media and Popular Imagination: Launching Radio, Television, and Digital Media in the United States*. Oxford: Oxford UP, 2004. Print.

Bogost, Ian. *Alien Phenomenology: Or What It's Like to Be a Thing*. Minneapolis: U of Minnesota P, 2012. Print.

Bolter, Jay David. *Writing Space: The Computer, Hypertext, and the History of Writing*. Hillsdale, NJ: Lawrence Erlbaum Associates, 1991. Print.

Boon, Louis. "Variation and Selection: Scientific Progress Without Rationality." *Evolutionary Epistemology: A Multiparadigm Program*. Eds Werner Callebut and Rix Pinxten. Dordrecht: D. Reidel Publishing Company, 1987. 159–178. Print.

Bordered, Serge. *L'Enigme des Enfants-loups*. Paris: Publibook, 2007. Print.

Borges. Jorge Luis. *Labyrinths*. London: Penguin, 1964. Print.

Borgman, Albert. *Technology and the Character of Contemporary Life*. Chicago: Chicago UP, 1987. Print.

Bourdain, Anthony. *Kitchen Confidential*. London: Bloomsbury, 2001. Print.

Brockman, John, ed. *Is The Internet Changing the Way You Think?* New York: Harper, 2011. Print.

Bruinsma, Max. "Watching, Formerly Reading." *I Read Where I Am*. Eds Mieke Gerritzen, Geert Lovink, and Minke Kampman. Valiz, 2011. Web. 15 July 2011.

Bryant, Levi. "Onticology– A Manifesto for Object-Oriented Ontology Part I." *Larval Subjects*. Ed. Levi Bryant. N.p., 12 Jan. 2010. Web. 15 Sept 2014.

Bulliet, Richard. "Determinism and Pre-Industrial Technology." *Does Technology Drive History? The Dilemma of Technological Determinism*. Eds. Merritt Roe Smith and Leo Marx. Cambridge (Mass.): MIT Press, 1994. 201–216. Print.

Bush, Vannevar. "As We May Think." *Atlantic Monthly* 176 (1945): 101–108. Rpt. in *The New Media Reader*. Eds Noah Wardrip-Fruin and Nick Montfort. Cambridge (Mass.): MIT Press, 2003. 37–47. Print.

Cairns, Douglas. *A History of Distributed Cognition*. University of Edinburgh, 2014. Web. 30 September 2014.

Callebut, Werner and R. Pinxten, eds. *Evolutionary Epistemology: A Multiparadigm Program with a Complete Evolutionary Epistemology Bibliography*. Dordrecht: D. Reidel, 1987. Print.

Calvino, Italo. *If on a Winter's Night a Traveller*. New York: Everyman's Library, 1993. Print.

Campana, Ellen "With a Wave of the Hand." *Scientific American* 19 May 2009. Web. 1 July 2009.

Campbell, Donald T. "Selection Theory and the Sociology of Scientific Validity." *Evolutionary Epistemology: A Multiparadigm Program*. Eds Werner Callebut and Rix Pinxten. Dordrecht: D. Reidel Publishing Company, 1987. 139–158. Print.

Carlson, T. A., G. Alvarez, D. A. Wu, and F. A.Verstraten. "Rapid Assimilation of External Objects into the Body Schema." *Psychological Science* 21.7 (2010): 1000–1005. Print.

Carmody, Tim. "Immanence and Transcendence in New Media." *Book Futurism* 25 March 2010. Web. 19 February 2011.

———. Blog comment. "Reports on the Changing Bodies of Books." *4oh4 – words not found* 23 March 2011. Web. 23 March 2011.

Carr, Nicholas. "Is Google Making Us Stupid?" *The Atlantic*. 1 July 2008. Web. 8 Sept 2014.

———. *The Shallows*. London: Atlantic Books, 2011. Print.

Chiel, H. J. and R. D. Beer. "The Brain Has a Body: Adaptive Behavior Emerges from Interactions of Nervous System, Body and Environment." *Trends In Neurosciences* 20 (1997): 553–557. Print.
Chemero, Anthony. *Radical Embodied Cognitive Science*. Cambridge (Mass.): MIT Press, 2011. Print.
Chen, Y., M. Ding, and J. Kelso. "Origin of Timing Errors in Human Sensorimotor Coordination." *Journal of Motor Behavior* 33 (2003): 3–8. Print.
Cicconi, Sergio. "Hypertextuality." *Mediapolis*. Ed. Sam Inkinen. New York: De Gruyter, 1999. 21–43. Print.
Clark, Andy. "Folk Psychology, Thought, and Context." *Microcognition*. Cambridge (Mass.): MIT Press, 1989. 37–59. Print.
———. *Natural-Born Cyborgs*. Oxford: Oxford UP, 2003. Print.
———. *Supersizing the Mind*. Oxford: Oxford UP, 2008. Print.
———. *Being There: Putting Brain, Body and World Together Again*. Cambridge (Mass.): MIT Press, 1997. Print.
Clark, Andy and David J. Chalmers. "The Extended Mind." *Analysis* 58 (1998): 10–23. Print.
Connor, Steven. "What Is It That It Is?" Birkbeck U of London n.d. Web. 14 June 2011.
Coover, Robert. "The End of Books." *The New York Times Book Review* 11 (21 June 1992): 23–25. Rpt. in *The New Media Reader*. Eds Noah Wardrip-Fruin and Nick Montfort. Cambridge (Mass.): MIT Press, 2003. 706–709. Print.
Crain, Caleb. "Twilight of the Books." *New Yorker* 24 December 2007. Web. 15 May 2009.
Crawford, Matthew B. *Shop Class as Soul Craft: An Inquiry Into the Value of Work*. New York: Penguin, 2009. Print.
Damasio, Antonio. *Descartes' Error*. London: Vintage, 2006. Print.
David, Jennifer. "Color and Acuity Differences between Dogs and Humans." *University of Wisconsin*, 1998. Web. 5 May 2010.
Davis, Lennard. "The End of Identity Politics and the Beginning of Dismodernism: On Disability as an Unstable Category." *Bending Over Backwards: Disability, Dismodernism, and Other Difficult Positions*. New York: NYU Press, 2002: 9–32. Print.
Davoli, Christopher C., Feng Du, Juan Montana, Susan Garverick, and Richard A. Abrams. "When Meaning Matters, Look but Don't Touch: The Effects of Posture on Reading." *Memory and Cognition* 38.5 (2010): 555–562. Print.
Dawkins, Richard. "Universal Darwinism." *Evolution from Molecules to Man*. Ed. D. S. Bendall. Cambridge: Cambridge UP, 1983. Print.
———. *The Selfish Gene: 30th Anniversary Edition*. Oxford: Oxford UP, 2006. Print.
Dee, Tim. "Nature Writing." *Archipelago* 5 (2011): 20–30. Print.
Dehaene, Stanislas. *Reading in the Brain*. New York: Viking, 2009. Print.
Deleuze, Gilles and Félix Guattari. *A Thousand Plateaus: Capitalism and Schizophrenia Part II*. Trans. Brian Massumi. London: Continuum, 2009. Print.
Derrida, Jacques. *Glas*. Trans. John P. Leavey Jr. and Richard Rand. Lincoln, NE: U of Nebraska P, 1990. Print.
———. *Memoirs of the Blind*. Chicago: U of Chicago P, 1993. Print.
———. *Of Grammatology*. Trans. Gayatri Spivak. Baltimore: Johns Hopkins UP, 1998. Print.

———. "Living On/Border Lines." Trans. James Hulbert. *Deconstructionism and Criticism*. Ed. Harold Bloom. London: Routledge, 2004. 62–142. Print.

———. *Paper Machine*. Trans. Rachel Bowlby. Stanford, CA: Stanford UP, 2005. Print.

DeVore, Paul W. *Technology: An Introduction*. Worcester, (Mass.): Davis Publishers, 1980. Print.

Di Pino, Giovanni, Angelo Maravita, Loredana Zollo, Eugenio Guglielmelli, and Vincenzo Di Lazzaro. "Augmentation-Related Brain Plasticity." *Frontiers in Systems Neuroscience* 8 (2014). Web. 30 September 2014.

Dobbs, David. "Is Page Reading Different From Screen Reading?" *Wired*, 20 September 2010. Web. 26 September. 2010.

Donald, Merlin. *Origins of the Modern Mind: Three Stages in the Evolution of Culture and Cognition*. Cambridge (Mass.): Harvard UP, 1993. Print.

———. *A Mind so Rare*. New York: Norton, 2001. Print.

Donath, Judith. "1964 Ford Falcon." *Evocative Objects*. Ed. Sherry Turkle. Cambridge, (Mass.): MIT Press, 2007. 153–161. Print.

Dorfman, Anna. "Tree of Codes." *Door Sixteen* 18 November 2011. Web. 26 June 2011.

Dotov, Dobromir G., Lin Nie, and Anthony Chemero. "A Demonstration of the Transition from Ready-to-Hand to Unready-to-Hand." *PLoS ONE* 5.3 (2010). Web. 2 July 2010.

Dreyfus, Hubert L. *Being-in-the-World*. Cambridge (Mass.): MIT Press, 1991. Print.

Drucker, Johanna. *The Visible Word: Experimental Typography and Modern Art, 1909–1923*. Chicago: U of Chicago P, 1994. Print.

———. *The Century of Artist's Books*. New York: Granary Books, 2004. Print.

Ehrenreich, Ben. "The Death of the Book." *Los Angeles Review of Books* 18 April 2011. Web. 21 April 2011.

Ellul, Jacques. *The Technological Society*. Toronto: Vintage, 1973. Print.

Emerson, Ralph Waldo. *Complete Works*. RWE.org n.d. Web. 17 July 2010.

Enzensberger, Hans Magnus. "Constituents of a Theory of Media." *New Left Review* 64 (1970): 13–36. Print.

Fagioli, S., B. Hommel, and R. I. Schubotz. "Intentional Control of Attention: Action Planning Primes Action-Related Stimulus Dimensions." *Psychological Research* 71 (2007): 22–29. Print.

Feenberg, Andrew. "Critical Theory of Technology: An Overview." *Tailoring Biotechnologies* 1.1 (2005): 47–64. Print.

Ferro-Thomsen "Reading Beyond Words." *I Read Where I Am*. Eds. Mieke Gerritzen, Geert Lovink, and Minke Kampman. Valiz, 2011. Web. 15 July. 2011.

Finklestein, David and Alistair McCleery *The Book History Reader*. London: Routledge, 2006. Print.

Flood, Alison. "Making Scents out of Novels." *Guardian* 5 June 2009. Web. 16 January 2011.

Floridi, Luciano. *Information: A Very Short Introduction*. Oxford: Oxford UP, 2010. Print.

———. *The Philosophy of Information*. Oxford: OUP, 2011. Print.

Fodor, Jerry. *The Language of Thought*. Cambridge (Mass.): Harvard UP, 1975. Print.

Foer, Jonathan Safran. *Tree of Codes*. London: Visual Editions, 2010. Print.

Foucault, Michel. *Technologies of the Self: A Seminar with Michel Foucault*. Eds L. H. Martin, H. Gutman and P. H. Hutton. Amherst, MA: U of Massachusetts P, 1988. Print.

Fox, Nicols. *Against the Machine: The Hidden Luddite Tradition in Literature, Art, and Individual Lives*. Washington, DC: Island Press, 2002. Print.

Fradera, Alex. "Remembering Together – How Long-Term Couples Develop Interconnected Memory Systems." *BPS Research Digest*. Ed. Christian Jarrett. British Psychological Society, 29 July 2014. Web. 24 August 2014.

Freud, Sigmund. *The Uncanny*. Trans. David McLintock. London: Penguin, 2003. Print.

Friis, Jan Kyrre Berg Olsen, Stig Andur Pedersen, and Vincent F. Hendricks, eds. *A Companion to the Philosophy of Technology*. Malden: Blackwell, 2013. Print.

Frost, Gary. "Reading by Hand." *futureofthebook.com*. n.p. 14 July. 2007. Web. 20 August. 2007.

Futuyma, Douglas. *Evolution*. Sunderland (Mass.): Sinauer Associates, 2013. Print.

Gallagher, Shaun. *How the Body Shapes the Mind*. Oxford: Oxford UP, 2006. Print.

Gallagher, Shaun and Jonathan Cole. "Body Schema and Body Image in a Deafferented Subject." *Journal of Mind and Behaviour* 16 (1995): 369–390. Print.

Gallagher, Shaun and Dan Zahavi. *The Phenomenological Mind: An Introduction to Philosophy of Mind and Cognitive Science*. London: Routledge, 2008. Print.

Gelernter, David. "The Book Made Better." *New York Times* 14 October 2009. Web. 18 June 2010.

Genette, Gerard. *The Work of Art: Immanence and Transcendence*. Trans. G. M. Goshgarian. Ithaca: Cornell UP, 1997. Print.

Gettier, Edmund. "Is Justified True Belief Knowledge?" *Analysis* 23 (1963): 121–123. Print.

Gibson, James. *The Ecological Approach to Visual Perception*. Hillsdale, NJ: Lawrence Erlbaum Associates, 1986. Print.

Gibson, Kathleen and Tim Ingold. *Tools, Language, and Cognition in Human Evolution*. Cambridge: Cambridge UP, 1993. Print.

Gibson, William. *Idoru*. London: Penguin, 2006. Print.

Giddings, Seth and Martin Lister, eds. *The New Media and Technocultures Reader*. Abingdon: Routledge, 2011. Print.

Gittelman, Lisa. *Scripts, Grooves, and Writing Machines: Representing Technology in the Edison Era*. Stanford: Stanford UP, 1999. Print.

Gladwell, Malcom. *Outliers*. London: Penguin, 2008. Print.

Gleick, James. *The Information*. London: Fourth Estate, 2012. Print.

Godzinski Jr., Ronald. "(En)Framing Heidegger's Philosophy of Technology." *The Philosophy of Technology* 6.1 (2005). Web. 14 August 2010.

Goldin-Meadow, Susan. *Hearing Gesture: How Our Hands Help Us Think*. Cambridge (Mass.): Harvard UP, 2003. Print.

———. "Talking and Thinking With Our Hands." *Current Directions in Psychological Science* 15.1 (2006): 34–39. Print.

Goldin-Meadow, Susan, Wagner Cook, Susan, and Zachary A. Mitchell. "Gesturing Gives Children New Ideas About Math." *Psychological Science* 20.3 (2009): 267–272. DOI: 10.1111/j.1467-9280.2009.02297.x .

Goldsworthy, Andy. *Hand to Earth*. Eds Andy Goldsworthy and Terry Friedman. London: Thames and Hudson, 2011. Print.

Gopnik, Alison. "Incomprehensible Visitors from the Technological Future." *Is The Internet Changing the Way You Think?* Ed. John Brockman. New York: Harper, 2011, 271–274. Print.

Graham, Jorie. *Never*. Manchester: Carcanet 2002. Print.
Gray, John. *Straw Dogs: Thoughts on Humans and Other Animals*. London: Granta, 2002. Print.
Greenfield, Adam. "What Apple Needs to Do Now." *Speedbird* 25 June 2010. Web. 14 August 2010.
Greenfield, P. M. "Language, Tools and Brain: The Ontogeny and Phylogeny of Hierarchically Organized Sequential Behavior." *Behavioral and Brain Sciences* 14 (1991): 531–595. Print.
Greenfield, Susan. "We Are at Risk of Losing Our Imagination" *Guardian* 25 April 2006. Web. 19 June 2012.
Grossman, Lisa. "In the Blink of Bird's Eye, a Model for Quantum Navigation." *Wired* 27 January 2011. Web. 28 February 2011.
Hansen, Ronald and Maaike Froelich. "Defining Technology and Technological Education: A Crisis, or Cause for Celebration?" *International Journal of Technology and Design Education* 4.2 (1994): 179–207. Web. 18 September 2010.
Harman, Graham. *Tool-Being*. Peru, IL: Open Court, 2002. Print.
———. "Brief SR/OOO tutorial." *Object Oriented Philosophy* 23 July 2010. Web. 19 December 2010.
———. "Technology, Objects and Things in Heidegger." *Cambridge Journal of Economics* (2009): 1–9. Web. 14 May 2011.
———. "Quick Response to Shaviro." *Object-Oriented Philosophy*. Ed. Graham Harman. n.p., 7 May 2010. Web. 30 September 2014.
———. *The Quadruple Object*. Winchester: Zero Books, 2010. Print.
———. *Prince of Networks: Bruno Latour and Metaphysics*. Victoria: re.press, 2009. Print.
———. "The Road to Objects." *Continent* 1.3 (2011): 171–179. Web. 30 September 2014.
———. "Intentional Objects for Non-Humans." *Europhilosophie*. n.p., 2008. Web. 30 September 2014.
———. "The Transcript." *The Prince and the Wolf: Latour and Harman at the LSE*. Alresford: Zero Books, 2011. Print.
Harris, C., A. Barnier, J. Sutton, and P. Keil. "Couples as Socially Distributed Cognitive Systems: Remembering in Everyday Social and Material Contexts." *Memory Studies* 7 (2014): 285–297. Print.
Havelock, Eric A. *Origins of Western Literacy*. Toronto: Ontario Institute for Studies in Education, 1976. Print.
———. *Preface to Plato*. Cambridge, MA: Harvard UP, 1963. Print.
Hayler, Matthew. "Maybe the Dumbest Generation Came Before Us." *Teleread* 6 June 2008. Web. 30 September 2014.
———. *Incorporating Technology: A Phenomenological Approach to the Study of Artefacts and the Popular Resistance to E-reading*. University of Exeter. 10 November 2011. Web. 30 September 2014.
Hayles, N. Katherine. "Deeper into the Machine: The Future of Electronic Literature." *Culture Machine* 5 (2003). Web. 10 Nov. 2012.
———. *How We Became Posthuman*. Chicago: U of Chicago P, 1999. Print.
———. *Writing Machines*. Cambridge (Mass.): MIT Press, 2002. Print.
———. "Print Is Flat, Code Is Deep: The Importance of Media-Specific Analysis." *Poetics Today* 25.1 (2004): 67–90. Poetics Today. Duke UP, 2004. Web. 24 September 2010.

———. *Electronic Literature: New Horizons for the Literary*. Notre Dame, Indiana: U of Notre Dame P, 2008. Print.
———. *How We Think: Digital Media and Contemporary Technogenesis*. Chicago: U of Chicago P, 2012. Print.
Heidegger, Martin. *Being and Time*. Trans. John Macquarrie and Edward Robinson. Malden, (Mass.): Blackwell Publishing, 1962. Print.
———. *Discourse on Thinking*. Trans. John M. Anderson and E. Hans Freund. New York: Harper and Row, 1966. Print.
———. "The Thing." *Poetry, Language, Thought*. Trans. Albert Hoftstadter. London: Harper, 1975. Print.
———. "The Question Concerning Technology." *The Question Concerning Technology and Other Essays*. Trans. William Lovitt. New York: Garland Publishing, 1977. Print.
Henkel, Hubert. "Die Technik der Musikinstrumentenherstellung am Beispiel des Klassischen Instrumentariums." *Technik und Kunst*. Ed. Dietmar Guderian. Düsseldorf, 1994. 67–91. Print.
Heyser, Charles J. and Anthony Chemero. "Novel Object Exploration in Mice: Not All Objects Are Created Equal." *Behavioural Processes* 89.3 (2012): 232–238. March 2012. Web. 30 September 2014.
Hume, David. "Sceptical Doubts Concerning the Operations of the Understanding." *An Enquiry Concerning Human Understanding*. Oxford: Oxford UP, 2008. 18–29. Print.
Husserl, Edmund. *Logical Investigations vol. 2*. London: Routledge, 2001. Print.
———. *Ideas*. London: Routledge, 2002. Print.
———. *Cartesian Meditations*. Dordrecht: Kluwer, 1999. Print.
Hutchins, Edwin. *Cognition in the Wild*. Cambridge (Mass.): MIT Press. 1996. Print.
———. "How a Cockpit Remembers Its Speeds." *Cognitive Science* 19.3 (1995): 265–288. Print.
Huxley, J. S. "Guest Editorial: Evolution, Cultural and Biological." *Yearbook of Anthropology* (1955): 2–25. Print.
Ihde, Don. *Embodied Technics*. Automatic Press, 2010. Print.
———. *Experimental Phenomenology*. Second Edition. Albany: SUNY Press, 2012. Print.
———. *Postphenomenology and Technoscience: The Peking University Lectures*. New York: SUNY Press, 2009. Print.
———. *Technology and the Lifeworld*. Bloomington: U of Indiana P, 1990. Print.
Ingold, Tim. *The Perception of the Environment*. London: Routledge, 2000. Print.
"Internet Use 'Good for the Brain'." *BBC News*. BBC. 14 October 2008. Web. 15 October 2008.
Iriki, A., M. Tanaka, and Y. Iwamura. "Coding of Modified Body Schema during Tool Use by Macaque Postcentral Neurones." *Neuroreport* 7 (1996): 2325–2330. Print.
Ishibashi, H., S. Hihara, and A. Iriki. "Acquisition and Development of Monkey Tool Use: Behavioural and Kinematic Analyses." *Canadian Journal of Physiology and Pharmacology*. 78 (2000): 1–9. Print.
Jackson, Maggie. *Distracted: The Erosion of Attention and the Coming Dark Age*. New York: Prometheus, 2010. Print.

Jacoby, Susan. *The Age of American Unreason*. Brecon: Old Street Publishing, 2008. Print.
James, William. "Great Men, Great Thoughts, and the Environment." *The Atlantic Monthly* 46.276 (1880): 441–459. Print.
Johnson, B. S. *The Unfortunates*. New York: New Directions, 2008. Print.
Johnson, Steven A. *Everything Bad Is Good for You: Why Popular Culture Is Making Us Smarter*. London: Penguin, 2005. Print.
———. "Dawn of the Digital Natives – Is Reading Declining?" *Guardian* 2 July 2008. Web. 21 September 2011.
Jones, Steven E. *Against Technology: From the Luddites to Neo-Luddism*. New York: Routledge, 2006. Print.
Kaczynski, Theodore. *The Unabomber Manifesto, or Industrial Society and Its Future*. n.c.: Filiquarian Publishing, 2007. Print.
Kant, Immanuel. *Critique of Pure Reason*. London: Penguin, 2007. Print.
Kaufman, Alan. "The Electronic Book Burning." *Evergreen Review* 120 (2009). Web. 18 January 2010.
Kay, Alan and Adele Goldberg. "Personal Dynamic Media." *Computer* 10.3 (March 1977): 31–41. Rpt. in *The New Media Reader*. Eds Noah Wardrip-Fruin and Nick Montfort. Cambridge (Mass.): MIT Press, 2003. 393–404. Print.
Keen, Andrew. *The Cult of the Amateur*. Boston: Nicholas Brealey Publishing, 2008. Print.
Kelly, Kevin. *What Technology Wants*. New York: Penguin, 2010. Print.
Kempler, D. R. "Disorders of Language and Tool Use: Neurological and Cognitive Links" *Tools, Language and Cognition in Human Evolution*. Eds Kathleen R. Gibson and Tim Ingold. Cambridge: Cambridge UP. 193–215. Print.
Kirschenbaum, Matthew G. *Mechanisms: New Media and the Forensic Imagination*. Cambridge (Mass.): MIT Press, 2008. Print.
Kittler, Friedrich A. *Gramophone, Film, Typewriter*. Trans. Geoffrey Winthrop-Young and Michael Wurtz. Stanford: Stanford UP, 1999. Print.
Klein, Naomi. *The Shock Doctrine*. London: Penguin, 2008. Print.
Kress, G. *Literacy in the New Media Age*. London: Routledge, 2003. Print.
Lacan, Jacques. "The Mirror Stage as Formative of the Function of the I as Revealed in Psychoanalytic Experience." *Ecrits*. Trans. Bruce Fink. New York: Norton, 2006. 75–81. Print.
LaFrance, Adrienne. "In 1858, People Said the Telegraph Was 'Too Fast for the Truth'." *The Atlantic*. 28 July 2014. Web. 22 October 2014.
Lakoff, George and Mark Johnson. *Metaphors We Live By*. Chicago: U of Chicago P, 2003. Print.
———. *Philosophy in the Flesh*. New York: Basic, 1999. Print.
Landau A. N., L. Aziz-Zadeh, and R. B. Ivry. "The Influence of Language on Perception: Listening to Sentences About Faces Affects the Perception of Faces." *Journal of Neuroscience* 30 (2010): 15254–15261. Web. 30 September 2014.
Latour, Bruno. *An Enquiry into Modes of Existence*. Cambridge (Mass.): Harvard UP, 2013. Print.
———. "The Transcript." *The Prince and the Wolf: Latour and Harman at the LSE*. Alresford: Zero Books, 2011. Print.
———. "The Berlin Key." *Matter, Materiality and Modern Culture*. Ed. P. M. Graves-Brown. London: Routledge, 1991. 10–21. Print.

Latour, Bruno, Graham Harman, and Peter Erdélyi. *The Prince and the Wolf: Latour and Harman at the LSE*. Alresford: Zero Books, 2011. Print.
Leibniz, G. W. "Monadology." *Philosophical Essays*. Trans. Roger Ariew and Daniel Garber. Indianapolis: Hackett, 1989. Print.
Leith, Sam. "Is This the End for Books?" *Guardian*, 14 August 2011. Web. 17 August 2011.
Leroi-Gourhan, André. *L'homme et la Matière*. Paris: Albin Michel, 1943. Print.
———. *Milieu et Techniques*. Paris: Albin Michel, 1945. Print.
———. *Gesture and Speech*. Cambridge (Mass.): MIT Press, 1993. Print.
Levy, Steven. "The Future of Reading." *Newsweek*. 26 November 2007. 57–64. Print.
Littau, Karin. *Theories of Reading: Books, Bodies and Bibliomania*. Cambridge: Polity Press, 2008. Print.
Lupyan, Gary and Emily Ward. "Language Can Boost Otherwise Unseen Objects Into Visual Awareness." *Proceedings of the National Academy of Sciences of the USA*. 110 (2013): 14196–14201. Web. 30 September 2014.
Mackey, M. *Literacies Across Media: Playing the Text*. Vol. 2. London: Routledge, 2007. Print.
———. *Mapping Recreational Literacies: Contemporary Adults at Play*. New York: Peter Lang, 2007. Print.
Malafouris, Lambros. *How Things Shape the Mind*. Cambridge (Mass.): MIT, 2013. Print.
Man, John. *Alpha Beta*. London: John Wiley, 2006. Print.
Mangen, Anne. "Hypertext Fiction Reading: Haptics and Immersion." *Journal of Research in Reading* 31.4 (2008): 404–419. Print.
———. *The Impact of Digital Reading on Immersive Fiction Reading*. Saarbrücken: VDM Verlag, 2009. Print.
———. "Putting the Body Back into Reading." *Pædagogisk Neurovidenskab* (2013): 11–31. Print.
———. "Reading Linear Texts on Paper Versus Computer Screen." *International Journal of Educational Research* 58 (2013): 61–68. 5 January 2013. Web. 30 September 2014.
Mangen, Anne and Jean-Luc Velay. "Cognitive Implications of New Media." *Johns Hopkins Guide to Digital Media*. Eds. Marie-Laure Ryan et al. Baltimore: Johns Hopkins UP, 2014. Print.
Maravita, Angelo and Atsushi Iriki. "Tools for the Body (Schema)." *TRENDS in Cognitive Sciences* 8.2 (2004): 79–86. Print.
Marx, Karl. "Economic and Philosophical Manuscripts of 1844." *Karl Marx: Early Texts*. Ed. Dave McLellan. London: Blackwell, 1972. Print.
———. *Capital Volume 1*. London, Penguin, 1990. Print.
Marx, Leo. "The Idea of 'Technology' and Postmodern Pessimism." *Does Technology Drive History? The Dilemma of Technological Determinism*. Eds Merritt Roe Smith and Leo Marx. Cambridge (Mass.): MIT Press, 1994. 237–257. Print.
———. "Technology: The Emergence of a Hazardous Concept." *E-Technology and Culture* 51.3 (2010). Web. 14 June 2011.
Mary, Sebastian. "Will the Real iPod for Reading Stand Up Now Please?" *if:book*, 19 March 2009. Web. 1 November 2009.
McLuhan, Marshall. *The Medium is the Massage*. Harmondsworth: Penguin, 1964. Print.

———. "The Galaxy Reconfigured: or the Plight of Mass Man in an Individualist Society." *The Gutenberg Galaxy: The Making of Typographic Man.* Toronto: U of Toronto P, 1962. Rpt. in *The New Media Reader.* Eds Noah Wardrip-Fruin and Nick Montfort. Cambridge (Mass.): MIT Press, 2003. 194–202. Print.

———. "The Medium is the Message" *Understanding Media: The Extensions of Man.* New York: McGraw Hill, 1964. Rpt. in *The New Media Reader.* Eds Noah Wardrip-Fruin and Nick Montfort. Cambridge (Mass.): MIT Press, 2003. 203–209. Print.

———. *Understanding Media.* London: Routledge, 2005. Print.

McPherron, Shannon P., et al. "Evidence for Stone-Tool-Assisted Consumption of Animal Tissues Before 3.39 Million Years Ago at Dikika, Ethiopia." *Nature* 466 (2010): 857–860. Web. 18 June 2011.

Medawar, Peter. "Technology and Evolution." *CSCS* 26 June 1996. Web. 26 September 2010.

Menary, Richard, ed. *The Extended Mind.* Cambridge (Mass.): MIT Press, 2012. Print.

———. "Introduction to the Special Issue on 4E Cognition." *Phenomenology and the Cognitive Sciences* 9.4 (2010): 459–463. Print.

Merchant, G. "Writing the Future in the Digital Age." *Literacy* 41.3 (2007): 118–128. Print.

Merleau-Ponty, Maurice. *Phenomenology of Perception.* Trans. C. Smith. London: Routledge, 1962. Print.

Mesthene, Emmanuel. *Technological Change.* New York: Mentor, 1970. Print.

Metzinger, Thomas. "The Pre-Scientific Concept of a 'Soul': A Neurophenomenological Hypothesis About its Origin." Johannes Gutenberg- Universität. 2003. Web. 15 Mar. 2011.

Mitcham, Carl. *Thinking Through Technology: The Path Between Engineering and Philosophy.* Chicago: U of Chicago P, 1994. Print.

Molina, Brett and Veronica Bravo. "Jony Ive: The Man Behind Apple's Magic Curtain." *USA Today* 19 September 2013. Web. 30 September 2014.

Morton, Timothy. *Ecology Without Nature.* Harvard UP, 2009. Print.

Munroe, Randall. "The Simple Answers." *XKCD* 11 November 2013. Web. 30 September 2014.

Neary, Lynn. "How E-Books Will Change Reading And Writing." *NPR.* NPR, 30 Dec. 2009. Web. 18 Sept. 2012.

Nelson, Richard R. "On the Nature and Evolution of Human Know-how." *Technology, Institutions, and Economic Growth.* Cambridge (Mass.): Harvard UP, 2005. 115–138. Print.

Newton, Michael. *Savage Boys and Wild Girls: A History of Feral Children.* New York: Picador, 2004. Print.

Norman, Donald A. *The Design of Everyday Things.* Cambridge (Mass.): MIT Press, 1998. Print.

Novak, Joseph. "A Sense of Eidos." *EIDOS* 19.2 (2005): 1–6. Web. 4 April 2011.

Nie, Lin, Dobromir G. Dotov, and Anthony Chemero. "Readiness-to-hand, Extended Cognition, and Multifractality." *Proceedings of the 33rd Annual Meeting of the Cognitive Science Society.* 1835–1840.n.d. Web. 30 September 2014.

O'Donell, James. "My Fingers Have Become Part of My Brain." *Is The Internet Changing the Way You Think?* Ed. John Brockman. New York: Harper, 2011, 191–192. Print.

Ong, Walter J. *Orality and Literacy: The Technologizing of the Word.* London: Routledge, 1982. Print.

Oretga y Gasset, José. *Meditación de la técnica*. Madrid: Revista de Occidente, 1939. Print.
Pascual-Leone, Alvaro and Fernando Torres. "Plasticity of the Sensorimotor Cortex Representation of the Reading Finger in Braille Readers." *Brain* 116.1 (1993): 39–52. Print.
Passig, Katherine. "Commonplaces of Technological Critique." Trans Saul Lipetz. *Eurozine* 16 September 2010. Web. 14 December 2010.
Pierce, Charles Saunders. "The Fixation of Belief." *Popular Science Monthly* 12 (1877): 1–15. Print.
Peitsch, D., A. Fietz, H. Hertel, J. de Souza, D. F. Ventura, and R. Menzel. "The Spectral Input Systems of Hymenopteran Insects and Their Receptor-Based Colour Vision." *Journal of Comparative Physiology A: Neuroethology, Sensory, Neural, and Behavioral Physiology* 170 (1992): 23–40. Print.
Perdue, Peter C. "Technological Determinism in Agrarian Societies." *Does Technology Drive History? The Dilemma of Technological Determinism*. Eds Merritt Roe Smith and Leo Marx. Cambridge (Mass.): MIT Press. 1994, 169–199. Print.
Pinch, T. J. and Karin Bijsterveld. "'Should One Applaud?': Breaches and Boundaries in the Reception of New Technology in Music." *Technology and Culture* 44.3 (2003): 536–559. John Hopkins UP. Web. 1 August 2014.
Pinker, Steven. *The Language Instinct*. London: Penguin, 1995. Print.
Pitt, Joseph C. *Thinking About Technology: Foundations of the Philosophy of Technology*. New York: Seven Bridges Press, 2000. Print.
Plato. *Plato in English*. Trans. H. N. Fowler. London: William Heinemann, 1919. Print.
Plotkin, Henry. *Darwin Machines and the Nature of Knowledge*. London: Penguin, 1995. Print.
Polastron, Lucien X. *The Great Digitization and the Quest to Know Everything*. Trans. Jon E. Graham. Rochester, Vermont: Inner Traditions International, 2006. Print.
Popper, Karl R. *Objective Knowledge: An Evolutionary Approach*. 1972. Oxford: Oxford UP, 1979. Print.
Preston, Beth. "Cognition and Tool Use." *Mind & Language* 13.4 (1998): 513–547. Print.
Radman, Zdravko, ed. *The Hand, an Organ of the Mind: What the Manual Tells the Mental*. Cambridge, MA: MIT, 2013. Print.
Reed, Catherine L., Jefferson D. Grubb, and Cleophus Steele. "Hands Up: Attentional Prioritization of Space Near the Hand." *Journal of Experimental Psychology* 32.1 (2006): 166–177. Print.
Reed, Catherine L., Ryan Betz, John P. Garza, and Ralph J. Roberts, Jr. "Grab It! Biased Attention in Functional Hand and Tool Space." *Attention, Perception, and Psychophysics* 72.1 (2010): 236–245. Print.
Richmond, Shane. "The Printed Book Is Doomed: Here's Why" *Telegraph*, 4 August 2011. Web. 18 August 2011.
Ridley, Mark. *Evolution*. Oxford: Blackwell Science, 2004. Print.
Rochat, Philippe. "Self-Perception and Action in Infancy." *Experimental Brain Research* 123 (1998): 102–109. Web. 30 September 2014.
Roome, Christine Shaw. "I've Got the Screen Eyes to Prove It: How Do Ebooks Really Compare to Traditional Books." *Life as a Human* 27 February 2011. Web. 15 March 2011.

Rosen, Larry. iDisorder: Understanding Our Obsession with Technology and Overcoming Its Hold on Us. New York: Palgrave, 2012. Print.

Rothenberg, David. Hand's End: Technology and the Limits of Nature. Berkley: University of California Press, 1993. Print.

Rowlands, Mark. The New Science of Mind. Cambridge (Mass.): MIT Press, 2010. Print.

Ruxin, Mark. "The Death of Touch and the Lost Joy of the Unexpected" *Huffington Post*, 29 June 2010. Web. 16 Jauary 2010.

Rymer, Russ. *Genie: A Scientific Tragedy*. New York: Harper Collins, 1994. Print.

Scharff, Robert C. and Val Dusek, eds. *Philosophy of Technology: The Technological Condition – An Anthology*. Malden, MA: Wiley-Blackwell, 2002. Print.

Scheinost, D., T. Stoica, J. Saksa, X. Papademetris, R. T. Constable, C. Pittenger, and M. Hampson. "Orbitofrontal Cortex Neurofeedback Produces Lasting Changes in Contamination Anxiety and Resting-State Connectivity." *Translational Psychiatry* 30 April 2013. Web. 30 September 2014.

Schendel, K. and L. C. Robertson. "Reaching Out to See: Arm Position Can Attenuate Human Visual Loss." *Journal of Cognitive Neuroscience* 16 (2004): 935–943. Print.

Schofield, Sarah. "Ten Things I Hate About eReaders." *I Call It Research* 24 February 2011. Web. 26 February 2011.

Shannon, Claude. "A Mathematical Theory of Communication." *The Bell System Technical Journal* 27 (1948): 5–83. Print.

Shapiro, Lawrence. *Embodied Cognition*. Abingdon: Routledge, 2010. Print.

Shennan, Steve, ed. *Pattern and Process in Cultural Evolution* Ewing, NJ: U of California P, 2009. Print.

Siegel, Lee. *Against the Machine: Being Human in the Age of the Electronic Mob*. London: Serpent's Tail, 2008. Print.

Skrupskelis, Ignas K. "Evolution and Pragmatism: An Unpublished Letter of William James." *Transactions of the Charles S. Peirce Society: A Quarterly Journal in American Philosophy* 43.4 (2007): 745–752. Project MUSE. Web. 8 August 2014.

Small, Gary. *iBrain: Surviving the Technological Alteration of the Modern Mind*. London: Harper, 2009. Print.

Small, G. W., T. D. Moody, P. Siddarth, and S. Y. Bookheimer. "Your Brain on Google: Patterns of Cerebral Activation During Internet Searching." *American Journal of Geriatric Psychiatry* 17 (2009): 116–126. Print.

Smell of Books. Durosport Electronics n.d. Web. 14 May 2011.

Smith, Merritt Roe and Leo Marx, eds. *Does Technology Drive History? The Dilemma of Technological Determinism*. Cambridge (Mass.): MIT Press, 1994. Print.

Sontag, Susan. *On Photography*. New York: Farrar, Straus & Giroux, 1977. Print.

Sorrell, Charlie. "New Book Smell: The Smell of Books in a Spray-Can." *Wired* 9 June 2009. Web. 16 January. 2011.

Stiegler, Bernard. *Technics and Time, 1: The Fault of Epimetheus*. Trans. Richard Beardsworth and George Collins. Stanford: Stanford UP, 1998. Print.

Stroop, J. R. "Studies of Interference in Serial Verbal Reactions." *Journal of Experimental Psychology* 18 (1935): 643–662. Print.

Swoyer, Chris. "How Does Language Affect Thought?" *University of Oklahoma*. University of Oklahoma, 14 April 2010. Web. 30 September 2014.

Symes, Ed, Giovanni Ottoboni, Mike Tucker, Rob Ellis, and Alessia Tessari. "When Motor Attention Improves Selective Attention: The Dissociating Role of Saliency." *The Quarterly Journal of Experimental Psychology* 63.7 (2010): 1387–1397. Web. 24 February 2011.

Symes, Ed, Mike Tucker, and Giovanni Ottoboni. "Integrating Action and Language through Biased Competition." *Frontiers in Neurobiology* 4 (2010): 1–13. Print.

Symes, Ed, Mike Tucker, Rob Ellis, Lari Vainio, and Giovanni Ottoboni. "Grasp Preparation Improves Change-Detection for Congruent Objects." *Journal of Experimental Psychology: Human Perception and Performance* 34.4 (2008): 854–871. Web. 15 January 2011.

Taylor, Timothy. *The Artificial Ape*. Basingstoke: Palgrave Macmillan, 2010. Print.

Thompson, Clive. "Retro Design is Crippling Innovation." *Wired.co.uk*. Ed. Dan Smith. Condé Nast, 7 Feb. 2012. Web. 14 Aug. 2013.

Thoreau, Henry David. *Walden*. Oxford: Oxford UP, 2008. Print.

Truss, Lynne. *Eats Shoots and Leaves*. London: Profile Books, 2003. Print.

Turkle, Sherry. *Life on the Screen: Identity in the Age of the Internet*. New York: Touchstone, 1995. Print.

———, ed. *Evocative Objects*. Cambridge (Mass.): MIT Press, 2007. Print.

Tweney, Dylan. "Why We Are Obsessed With the iPad." *Wired* 1 April 2010. Web. 1 April 2010.

Ulin, David L. *The Lost Art of Reading: Why Books Matter in a Distracted Time*. Seattle: Sasquatch, 2010. Print.

van Orden, G., J. Holden, and M. Turvey. "Self-Organization of Cognitive Performance." *Journal of Experimental Psychology*: General 132 (2003): 331–350. Print.

Varela, Francisco J., Evan Thompson, and Eleanor Rosch. *The Embodied Mind: Cognitive Science and Human Experience*. Cambridge (Mass.): MIT Press, 1993. Print.

Vehkavaara, Tommi. "Extended Concept of Knowledge for Evolutionary Epistemology and for Biosemiotics: Hierarchies of Storage and the Subject of Knowledge." *Emergence, Complexity, Hierarchy, Organization* 91.8 (1998): 207–216. Print.

Vendler, Helen. "Jorie Graham: The Moment of Excess." *Jorie Graham: Essays on the Poetry*. Ed. Thomas Gardner. Wisconsin: U of Wisconsin P, 2005. Print.

Verbeek, Peter-Paul. *What Things Do*. University Park: U of Pennsylvania P, 2005. Print.

Verdoux, Philippe. "Human Evolution and Technology: From Prosimian to Posthuman." *IEET* 16 April 2010. Web. 15 August 2010.

Vershbow, Ben. "The Cramped Root: Worshipping the Artifact." *if:Book* June 2005. Web. 14 March 2010.

Vezina, Danielle Suzanne. *Phenomenology and Dance*. Diss. Simon Fraser U, 2006. Print.

Vincenti, Walter G. *What Engineers Know and How They Know It*. Baltimore: Johns Hopkins UP, 1990. Print.

Visel, Dan. "Learning to Read." *if:Book* 20 April 2006. Web. 18 March 2010.

Vygotsky, Lev. *Mind in Society: The Development of Higher Psychological Processes*. Cambridge (Mass.): Harvard UP, 1978. Print.

———. "The Instrumental Method in Psychology." *The Collected Works of L. S. Vygotsky: Volume 1 – Problems of the Theory and History of Psychology*. Eds Robert Rieber and Jeffery Wollock. New York: Plenum Press, 1997. Print. 85–89.

Wallace, David F. "Federer as Religious Experience." *New York Times* 20 Aug. 2006. Web. 4 Oct. 2012.

Walsh, M. "The 'Textual Shift': Examining the Reading Process with Print, Visual and Multimodal Texts." *Australian Journal of Language and Literacy* 29.1 (2006): 24–37. Print.

Walsh, M., J. Asha, and N. Sprainger. "Reading Digital Texts." *Australian Journal of Language and Literacy*. 30.1 (2007): 40–53. Print.

Ward, David and Mog Stapleton. "Es Are Good: Cognition as Enacted, Embodied, Embedded, Affective and Extended." *Consciousness in Interaction: The Role of the Natural and Social Context in Shaping Consciousness*. Amsterdam: John Benjamins, 2012. 89–104. Print.

Whitehead, Alfred North. *Process and Reality*. New York: Free Press, 1978. Print.

Williams, Rosalind. "The Political and Feminist Dimensions of Technological Determinism." *Does Technology Drive History? The Dilemma of Technological Determinism*. Eds. Merritt Roe Smith and Leo Marx. Massachusetts: MIT Press, (1994): 217–235. Print.

Williams, William Carlos. *In the American Grain*. New York: New Directions, 1956. Print.

Wilson, Margaret. "Six Views of Embodied Cognition." U of Wisconsin. n.d. Web. 23 March 2011.

Winner, Langdon. *Autonomous Technology: Technics-Out-of-Control as a Theme in Political Thought*. Cambridge (Mass.): MIT Press, 1978. Print.

Wolf, Maryanne. *Proust and the Squid*. Cambridge: Icon, 2008. Print.

Wuketits, Franz M., ed. *Concepts and Approaches in Evolutionary Epistemology: Towards and Evolutionary Theory of Knowledge*. Dordrecht: D. Reidel Publishing Company, 1984. Print.

Zahavi, Dan. *Husserl's Phenomenology*. Stanford: Stanford UP, 2003. Print.

Zerzan, John. *Future Primitive*. New York: Autonomedia, 1994. Print.

———. *Elements of Refusal*. Columbia: C.A.L., 1999. Print.

———. "Against Technology: A Talk by John Zerzan April 23, 1997." The Anarchist Library. 19 July 2009. Web. 26 July 2011.

———. *Running on Emptiness*. Los Angeles: Feral House, 2002. Print.

Ziman, John, ed. *Technological Innovation as an Evolutionary Process*. Cambridge: Cambridge UP, 2003. Print.

Zimmer, Carl. *Evolution: The Triumph of an Idea*. London: Arrow, 2003. Print.

Zubiri, Xavier. *On Essence*. Trans. A. R. Caponigri. Washington, DC: Catholic University Press, 1980. Print.

Index

Note: All authors indexed in bold.

affordances, 5, 144–7, 152, 154, 159, 169, 171, 173, 180, 194, 202, 209–10, *see also* Chemero, Anthony; James Gibson; radical embodied cognitive science

Birkerts, Sven, 1, 8–9, 36–7, 61, *see also* technology, resistance to
book (paper/print), 1, 3, 4, 8–9, 11, 19–29, 31–3, 36–40, 51, 117, 121, 123, 135, 137, 139–43, 147, 157, 186, 211–14, 224–5

Carr, Nicholas, 29–30, 33–4, 157
Chemero, Anthony, 48, 79, 99, 101, 144, 145, 148, 152–4, 161, 168–9, 171, 172, 173, 180, *see also* radical embodied cognitive science
Clark, Andy, 3, 48, 52, 71, 77, 80–1, 89, 93, 113, 196, *see also* cognition, embodied; cognition, extended
cognition
  distributed, 148, 149, 194–5, 197
  embedded, 148, 151–2
  embodied, 5, 45–56, 76–90, 98–103, 127, 129, 149–50, 152–3, 154–62
  enactive, 73, 148, 151–2
  extended, 76–90, 98–103
  situated, 73, 148, 151–2
cognitive science, 2, 5, 20, 30, 39, 37–8, 40, 47, 78–9, 82, 87, 99, 101, 121, 132, 144–5, 148–54, 161, 201
communality, 4, 58, 70–6, 84, 86, 89–90, 91, 96, 97, 108, 110, 112–13, 115–16, 130, 149, 160, 221, *see also* technology, new definition of

data, 6, 125, 127, 164, 166, 192–208, 210, 215, 218, 219–26, 231, *see also* information; knowledge

device, 5, 7, 75, 90, 94–103, 108–9, 112, 113, 117, 120, 129, 160, 225, 231–2, *see also* technology, new definition of
domestication, 5, 71, 91–4, 96–7, 106, 112–13, 108, 110, 115, 116, 160, 177, 191, *see also* technology, new definition of

e-reading
  e-book, 21, 25, 228
  e-reader, 1, 4, 11, 20–2, 25, 27, 29–30, 39, 56, 71, 76, 90, 94, 97, 112, 121, 134, 138, 144, 175–8, 181–3, 224, 225, 226–8, 232
  iPad, 1, 25, 29, 117, 127, 142, 228, 232
  Kindle, 1, 29, 143, 227, 228, 232
embodiment
  of artefact, 2, 5, 19, 20, 27, 144–7, 154–62, 211–15, 220–9
  of knowledge, 164, 193–206, 215–29
  of user, 2, 5, 19, 21, 47, 76–90, 124, 125–8, 147–8, 154–62
epistemology, *see* Knowledge
evolutionary epistemology, 6, 133, 166, 208, 214–29, *see also* Plotkin, Henry
expertise, 2, 5–6, 56, 68–9, 76–7, 79, 81–90, 97–8, 102–3, 112, 119, 120–2, 123, 125, 128–9, 133–4, 139, 154, 165–7, 169, 171, 172, 176–7, 179, 182, 184, 187–9, 193–5, 199, 200, 207, 229–31, 233, *see also* technology, new definition of
extension, 4, 37, 44, 46, 52, 68, 70, 71, 72, 76, 77, 79, 80, 81, 82, 86, 89, 90, 91, 95, 96, 108, 110, 113, 115, 116, 177, 194, 229, *see also* technology, new definition of

249

folk phenomenology, 3, 4, 19, 23–7, 32, 39, 51, 53, 55, 121, 137, 151, 190, 193, 225, *see also* phenomenology

Gallagher, Shaun, 20–1, 48, 74, 79, 114
gestalt, *see* whole-composite
Gibson, James, 144–5, 152, 154, 171, 173, 180, *see also* affordances; Chemero, Anthony, radical embodied cognitive science

haptics, *see* touch
Harman, Graham, 88, 120, 165–93, 202–3, 206, 207, *see also* object-oriented ontology, *see also* Heidegger, Martin
Hayles, N. Katherine, 20, 22, 27–8, 33–4, 37, 92, 134–5, 142
Heidegger, Martin, 16–17, 64–5, 67–8, 69, 77, 87–9, 98–103, 109, 113, 114, 170–2, 179, 181, 187, *see also* phenomenology
Husserl, Edmund, 21, 119, 126–8, 147–8, 167–8, 171–2, 179–80, 182–3, 187–9, *see also* phenomenology

Ihde, Don, 121, 126, 130–1, 155–6, 160, 198, 205, *see also* postphenomenology
incorporation, 4, 71, 76–90, 91, 92, 93, 95, 96, 97, 98–103, 106, 108, 110, 112, 113, 115, 116, 119, 149, 151, 155, 160, 170–2, 181–2, 184, 188, 193, 195, 208, 225, *see also* technology, new definition of
information, 6, 42, 121, 128, 164–6, 190, 192, 195, 197, 201–6, 207, 216, 217, 218–31, *see also* data; Knowledge
iPad, *see* e-reading

Kindle, *see* e-reading
Knowledge, 5–6, 23, 63, 64, 65, 70, 73, 121, 125, 131, 135–6, 144, 164–6, 175, 186–90, 191–206, 207–9, 230–4, *see also* data; information

Lakoff, George & Mark Johnson, 48, 124, 129, 133, 136–8, 140–1, *see also* metaphor
Latour, Bruno, 66, 162, 173–4, 176–8, *see also* Harman, Graham; object-oriented ontology
Luddite, 10–14, 18, 62–3, 93

McLuhan, Marshall, 71, 89, 108, 158, 214–15
Malafouris, Lambros, 148, 194–5, *see also* cognition
Mangen, Anne, 2, 3, 22, 29–30, 34
metaphor, 5, 39, 41, 120, 124, 129–44, 159, 165, 200, *see also* Lakoff, George & Mark Johnson

Neo-Luddite, *see* Luddite
Neuroscience, *see* cognitive science

object-oriented ontology, 2, 5, 88, 120, 163, 165–93, 201, 206, 207–8, 223, 227, *see also* Harman, Graham; Heidegger, Martin
Ong, Walter, 28, 42, 43, 58, 80, 106, 158–9

phenomenology, 16–17, 21, 24, 39, 47–8, 87–90, 98–103, 109–10, 113, 114, 119, 126–8, 130–1, 147–8, 162, 166–8, 170–2, 179–80, 181, 182–3, 187–9, *see also* Heidegger, Martin; Husserl, Edmund; postphenomenology
Pitt, Joseph C., 61–2, 66, 69–70, 198
Plotkin, Henry, 133, 209, 216–23, *see also* evolutionary epistemology; Knowledge; technology, as something evolved
postphenomenology, 2, 5, 121, 126, 130–1, 144, 148, 153, 154–63, 160, 168–9, 171, 173–4, 177, 184, 192, 194, 198, 205, *see also* Ihde, Don; Verbeek, Peter-Paul

radical embodied cognitive science, 48, 79, 99, 101, 121, 145, 148, 152–4, 155, 161, 168, 169, 171, 172, 173, 194, 195, *see also* Chemero, Anthony; cognition readiness-to-hand, *see* Harman, Graham; Heidegger, Martin; incorporation

Taylor, Timothy, 16–17, 45–7, 95–6, 111–12, 141–2, 157
technology
  existing definitions of, 3–4, 8, 44, 57, 60–9
  new definition of, 3–4, 69–118, 121, 154, 158, 198, 207, 225, *see also* communality; domestication; extension; incorporation
  resistance to, 1, 3–4, 6, 7–35, 37, 56–8, 62, 139, 185–6
  as something evolved, 6, 43, 49, 70, 140–1, 166, 211–15, 220–9, 232
  technique, 64–5, 72, 76, 86, 91, 108, 109, 205

technological system, 64–6, 72, 93, 117, 123, 149, 151

Verbeek, Peter-Paul, 154, 156, 161–3, 168–9, 171, 173–4, 177, 184, *see also* Ihde, Don; postphenomenology

whole-composite, 5, 72, 119–21, 123–47, 154–5, 160–2, 165, 172–3, 175, 178–80, 183, 184, 186, 188, 189, 190, 193, 194, 202, 204–6, 228

Zahavi, Dan, 20–1, 74, 127, 168, 189, *see also* Gallagher, Shaun; Husserl, Edmund; phenomenology

GPSR Compliance
The European Union's (EU) General Product Safety Regulation (GPSR) is a set of rules that requires consumer products to be safe and our obligations to ensure this.

If you have any concerns about our products, you can contact us on

ProductSafety@springernature.com

In case Publisher is established outside the EU, the EU authorized representative is:

Springer Nature Customer Service Center GmbH
Europaplatz 3
69115 Heidelberg, Germany

www.ingramcontent.com/pod-product-compliance
Lightning Source LLC
Chambersburg PA
CBHW071615100426
42873CB00004B/49